LEAN FOR SYSTEMS ENGINEERING WITH LEAN ENABLERS FOR SYSTEMS ENGINEERING

WILEY SERIES IN SYSTEMS ENGINEERING AND MANAGEMENT

Andrew P. Sage, Editor

A complete list of the titles in this series appears at the end of this volume.

LEAN FOR SYSTEMS ENGINEERING WITH LEAN ENABLERS FOR SYSTEMS ENGINEERING

Bohdan W. Oppenheim

A JOHN WILEY & SONS, INC., PUBLICATION

Published by John Wiley & Sons, Inc., Hoboken, New Jersey.
Published simultaneously in Canada.

For general information on our other products and services or for technical support, please contact our Customer Care Department within the United States at (800) 762-2974, outside the United States at (317) 572-3993 or fax (317) 572-4002.

Wiley also publishes its books in a variety of electronic formats. Some content that appears in print may not be available in electronic formats. For more information about Wiley products, visit our web site at www.wiley.com.

Library of Congress Cataloging-in-Publication Data:
Oppenheim, Bohdan W., 1948-
  Lean for systems engineering with lean enablers for systems engineering /
Bohdan W. Oppenheim.
    p. cm. – (Wiley series in systems engineering and management ; 82)
  Includes bibliographical references.
  ISBN 978-1-118-00889-8 (hardback)
  1. Engineering economy. 2. Systems engineering. 3. Lean manufacturing.
I. Title.
  TA177.4.O66 2011
  658.15–dc22

                        2010053404

Printed in Singapore

oBook ISBN: 978-1-118-06399-6
ePDF ISBN: 978-1-118-06397-2
ePub ISBN: 978-1-118-06398-9

10 9 8 7 6 5 4 3 2 1

*To my sons Peter and Tomas, and to my beloved lands:
United States and Poland— I dedicate this book in the
hope that all four will always strive to optimize in the
next larger context without sub - optimization,
consistent with Lean Thinking.*

Contents

Foreword

Poor program performance for large complex systems is an all too frequent outcome in defense, aerospace, civil infrastructure, shipbuilding and many other domains. Although many programs are executed effectively, those that aren't slip months to years to decades behind schedule, overrun budgets by millions or billions of dollars, and become public black eyes. Often there are multiple interacting causes for poor program performance—for example change of key leaders, optimistic technology forecasts, inexperienced workforce, or politics. Yet study after study concludes that ineffective system engineering is a leading or contributing cause for almost every poorly performing program. For the past several decades there have been multiple and constant pleas to improve effectiveness of system engineering.

What have been the responses and how effective have they been? Universities, industry, and military organizations have launched system engineering training courses, undergraduate and graduate degrees, and certification programs. The International Council on Systems Engineering (INCOSE) has become established as the professional society for systems engineering. Its many working groups are communities of practice for systems engineers. Organizations have established high level system engineer leadership positions to bring management attention and priority to this aspect of their operations. Research efforts have been funded to devise new approaches or better understand the practice of systems engineering. New tools and metaphors for system engineering are being developed: Model-Based System Engineering, modeling languages, and requirements management software tools are three examples. A newly emerging approach is the Meta program which will utilize a metalanguage with an aim to "enable 'fabless' development of systems that are probabilistically verifiable as correct by design, compressing the cycle time by removing manufacturing from the loop",[1] Although each of these responses have had or will have a positive impact, none of them individually or all of them collectively seem to have overcome the ineffectiveness of systems engineering in many applications.

This book presents a new response to the need to improve system engineering effectiveness. It is a response that supports and enables established, proven systems engineering methods to become more effective. It is a response in harmony with others, not in competition. It is a response based upon the knowledge base and proven effectiveness of lean six sigma thinking in many other functional domains. Lean which originated at Toyota is not a set of tools but a way of thinking. Lean Systems Engineering is not some kind of new fangled systems engineering. It is a way to make the best systems engineering better by applying common sense continuous process improvement approaches.

Lean for Systems Engineering is the first major work in the emerging field of lean systems engineering. The author lays out the fundamentals of Lean Thinking and how it applies to system engineering. The book is rich with citations to classic and recent literature, and provides a scholarly basis for the main product that is introduced - Lean Enablers for System Engineering (LEfSE). LEfSE is a compilation of best practices or "dos and don'ts" to supplement established system engineering practices. A major portion of the book is devoted to descriptive and useful information regarding the LEfSE and how to adopt and apply them. Although new and not yet widely deployed, the book presents encouraging results from early adopters of LEfSE. The implementation cost is very modest and the gains are significant: return on investment of 5% and reduction of waste or cycle time of 20–40%.

The Lean Enablers for System Engineering product is an outgrowth of the INCOSE Lean Systems Engineering Working Group. I was fortunate to be an early participant in this working group and play some role in the formation of LEfSE. But I was also wise enough to stand aside and let Bo Oppenheim's passion, experience and abilities to get things done lead this group to its successful outcome. I am very pleased to see this effort culminate in this new book in which Bo lays out the fundamentals of lean systems engineering, the history of the LEfSE development, and the details of this new product.

Lean for System Engineering is targeted at the practitioner who is trying to make systems engineering more effective in her or his organization or program. Yet its scholarly underpinnings make the text very suitable for teachers. Educators and trainers who wish to weave lean thinking into their systems engineering curriculum will find this an invaluable text.

I encourage readers who find value in lean for systems engineering to become involved in the INCOSE Lean Systems Engineering Working Group. It would be

---

[1] Aviation Week and Space Technology, Nov 8, 2010, pp 72–75.

especially valuable for you to share your experiences in applying or teaching Lean Enablers for System Engineering so that this ongoing effort can continue to become better and better.

EARLL M. MURMAN
MIT Ford Professor of Engineering Emeritus
Port Townsend, WA

Preface

This book was written for two purposes—to popularize the important emerging field termed *Lean Systems Engineering*, and to serve as a reference text for the first major product created under it: *Lean Enablers for Systems Engineering (LEfSE)*.

Balance between technical and business success is one of the critical aspects of high-quality Systems Engineering. Regretfully, during the past decade, in some government programs inadequate incentives tended to throw the balance off. Namely, system or mission assurance was incentivized at the expense of budget and schedule. This book describes how *Lean Thinking* is applied to Systems Engineering to achieve both technical and business successes.

Lean Thinking, or simply *Lean*, has been applied in many domains, including manufacturing, supply chain management, product development, administration, accounting, healthcare, and government, in each case creating significant, well-documented benefits. *Lean* dramatically reduced car development and production times and costs while increasing product quality and stakeholder satisfaction, and became a standard practice in most manufacturing industries.

Lean Systems Engineering should be regarded as the process of amending the well-established and sound traditional Systems Engineering process with the wisdom of *Lean Thinking*, rather than replacing Systems Engineering with a new body of knowledge. Most emphatically, *Lean Systems Engineering* does not mean "less systems engineering," but rather better Systems Engineering, with better preparations of the enterprise processes, people, and tools; better program planning and frontloading; better workflow management; and better program management and leadership with higher levels of responsibility, authority, and accountability. Under the *Lean Systems Engineering* philosophy, system success or mission assurance is non-negotiable, and any task which is legitimately needed for success must be included in the program. The benefits of *Lean Systems Engineering* are immediately visible in better Systems Engineering processes, as well as in the streamlined execution of the program, with less waste, waiting, and rework; more predictable and robust program flow; lower overall program cost and shorter schedule, and creation of better value to the customer. The fundamental feature of *Lean Systems Engineering* is to perform all preparations and planning of the people, processes, tools, and individual tasks well enough so that the *right* tasks can be executed *right the first time*, creating customer value while minimizing waste.

Lean Enablers for Systems Engineering (LEfSE) is the first major practical product created under Lean Systems Engineering and is the main topic of this book. *LEfSE* is a checklist of 147 practices formulated as dos and don'ts of Systems Engineering and some closely related aspects of Product Development (PD) and Enterprise Management, including supply chain management. The intent of *LEfSE* is to offer comprehensive and actionable practices to the profession, with the objective of improving overall performance of Systems Engineering and the PD program it serves; strengthen the value created by the program; increase the level of satisfaction of stakeholders; and reduce waste, program cost, and time. Most of the effort in developing Lean Systems Engineering and the *LEfSE* was conducted by the Lean Systems Engineering Working Group (LSE WG) of INCOSE. This was an intensive and rigorous two-year process involving 14 experts from industry, academia, and government, supported by members of the WG. Version 1.0 of *LEfSE* was released to INCOSE and to the public in January 2009. In 2010, the LEfSE were recognized with two following prestigious awards:

- The INCOSE Product of the Year
- The Shingo Award for Research and Publication

These independent recognitions from both *Systems Engineering* and *Lean* communities are regarded as an important step in validating *Lean Enablers for Systems Engineering*.

After the *LEfSE* were released to the public, the author and several of his colleagues on the development team were invited to offer numerous (40 at the time of this writing) workshops, tutorials, webinars, and seminars to industry, government, INCOSE meetings, and universities in the U.S., China, Finland, France, Italy, Israel, Netherlands, Norway, Poland, Singapore, Sweden, and the United Kingdom. Over 2000 participants took part in these professional events, generally excellently received. In many of these sessions participants asked for more explicit information about the enablers than was possible to present in the short sessions. This book is a direct response to these requests.

Both SE and Lean are a mix of art and science, and the art of effective teaming plays a critical role in SE. Some programs can be compared to the effort of climbing Mt. Everest or competitive racing in a storm—extraordinary teams pursuing difficult challenges against heavy odds. The Apollo, U2, F117, Boeing 747, nuclear submarine, I products from Apple, and the Prius car programs come to mind as examples. Such programs inspire young people to choose a career in engineering. The best programs require leadership, teamwork, enthusiasm, passion, inner energy, and joy, in addition to competence and experience. These elements have a prominent place in the book.

The material presented in the book is broad in scope, non-mathematical, and it applies to the development of all types of complex systems in all domains, both in commercial and government programs. We hope that LEfSE will benefit Systems Engineers and other Product Development engineers and managers, civilian and military professionals involved in systems acquisition both in industry and government, project managers, and faculty and students of these fields.

BOHDAN W. OPPENHEIM
Santa Monica, CA
May 3, 2011

Acknowledgments

Many people contributed to the development of knowledge captured in this book. The International Council on Systems Engineering (INCOSE), a most friendly and supportive professional society of Systems Engineers, served as a hospitable home for most of the activities described in the book. INCOSE enabled the project with workshops, symposia, web space, and friendly administrative support, and recognized the work with the INCOSE 2009 Best Product Award.

The Lean Systems Engineering Working Group (LSE WG) of INCOSE was initiated to develop and mature the knowledge of Lean Systems Engineering. Many members of the WG, too numerous to list by name (the current membership of 200 individuals is available on the INCOSE site), supported the development of 147 practices we term *Lean Enablers for Systems Engineering (LEfSE)* which are the main subject of this book. LEfSE were developed by the two following teams organized under the LSE WG:

Concept-to-Beta Team: Dr. Earll Murman, Ford Professor of Engineering Emeritus, Aeronautics and Astronautics, Massachusetts Institute of Technology, team leader; Col. James Horejsi, Chief Engineer, U.S. Space and Missile Command, ret.; James Zehmer, Vice President, Toyota ABC, Long Beach, California; Dr. Larry Earnest, Project Manager and lead Systems Engineer, Northrop Grumman; Michael Schaviatello, Project Manager, Boeing Satellite Development Co.; Deborah Secor and Ray Jorgensen, Systems Engineering Managers, Rockwell Collins; and the author, Bohdan ("Bo") Oppenheim, Professor of Systems Engineering, Loyola Marymount University.

Prototype Team: Dr. Larry Earnest (vide); Raymond Jorgensen (vide), Ron Lyells, Honeywell; Bohdan Oppenheim (vide), team leader; Uzi Orion, ELOP,

Israel; David Ratzer, Rockwell Collins; Deborah Secor (vide); Dr. Hillary G. Sil-litto, UK MoD Abbey Wood; Dr. Stan Weiss, Consulting Professor of Aeronautics and Astronautics and Director of Systems Engineering, Stanford University; and Dr. Avigdor Zonnenshain, Rafael, Israel.

Each enabler was originally formulated as a crisp sentence. In this book, I expanded each enabler into a page-long table, adding explanations, interpretations, implementation suggestions, lagging factors, and recommended reading lists, and presented the tables in Chapter 7. I alone take responsibility for all these interpretations.

Earlier this year, my colleagues and I published a full-length paper [Oppenheim, Murman, Secor, 2010] describing the development of LEfSE. Chapters 2, 3, 4.1, 5, and 6 of this book contain several long passages adapted from the article. I adopted Chapter 4.2 from my earlier article [Oppenheim, 2004].

My special gratitude goes to Professor Earll Murman and Deborah Secor, co-authors of the LEfSE paper, great friends, colleagues, and extraordinary leaders. Earll Murman initiated the LEfSE project and led the Beta Team. Deb Secor co-led the project from the beginning and has served as co-Chair and extraordinary co-leader of the INCOSE Lean SE Working Group. They generously contributed their time, leadership, wisdom, expertise, experience, and enthusiasm to the development of Lean Systems Engineering and Lean Enablers for Systems Engineering in critical ways that help make it a success. I am very grateful to Deb for reviewing the tables in Chapter 7 and making valuable suggestions.

Dr. Donna Rhodes, MIT, one of the early pioneers of Lean Systems Engineering, contributed meritorious comments and authored or supervised numerous research projects from which this project benefited. Drs. Eric Rebentisch and Hugh McManus from the MIT LAI and Dr. Stan Weiss of Stanford University led research in many important Lean areas used throughout this project.

I am grateful to Mr. Jeff Doyle and his Systems Engineering team at Northrop Grumman for sharing their practical experience with me.

Gratitude is due to all individuals, too numerous to list by name, who participated in the lengthy surveys critically important in the LEfSE project.

Paulos Ashebir, Jeffrey Dorey, and Eugene Plotkin, Mechanical Engineering and Systems Engineering graduate students at Loyola Marymount University, served as the capable and enthusiastic Research Assistants on the LEfSE project. The mapping of the enablers onto the SE processes listed in Appendix 2 was performed by Dana Makiewicz as a part of his MS capstone project. The mapping was requested by INCOSE, with the intent of including it into a future edition of the INCOSE Systems Engineering Handbook, but the organization has not yet approved the text included in Appendix 2. Eugene Plotkin and Jeff Dorey created Figure 6.1. Most of the figures have been redrawn to publication standards by Jeff Dorey. I am grateful to Paulos Ashebir for selecting the quotes from literature, which are included in many enabler tables.

I am immensely grateful to all my graduate students at Loyola Marymount University, too numerous to mention by name, and to fellow faculty who brought invaluable real-life Systems Engineering examples to my attention, which stimulated many ideas presented in this work.

I am grateful to my son Peter W. Oppenheim for his help in proofreading the entire manuscript. Any remaining imperfections are my fault and not his.

I am very grateful to the wonderful production team at John Wiley & Sons: Kuzhali of Laserwords, George Telecki, Editor, and Kellsee Chu, Production Editor.

List of Enablers and Subenablers in Chapter 7

Lean Principle 1: Value

1.1 Follow All Practices for the Capture and Development of Requirements, as Found in the INCOSE Handbook

1.2 Establish the Value of the End Product or System to the Customer

    1.2.1 Define value as the outcome of an activity that satisfies at least three conditions

    1.2.2 Define value-added in terms of value to the customer and his needs

    1.2.3 Develop a robust process to capture, develop, and disseminate customer value with extreme clarity

    1.2.4 Develop an agile process to anticipate, accommodate, and communicate changing customer requirements

    1.2.5 Do not ignore potential conflicts with other stakeholder values, and seek consensus

    1.2.6 Explain customer culture to program employees, i.e. the value system, approach, attitude, expectations, and issues

1.3 Frequently Involve the Customer

    1.3.1 Everyone involved in the program must have a customer-first spirit

    1.3.2 Establish frequent and effective interaction with internal and external customers

    1.3.3 Pursue an architecture that captures customer requirements clearly and can be adaptive to changes

2.3 Plan for Frontloading the Program

2.3.1 Plan to utilize cross-functional teams made up of the most experienced and compatible people at the start of the project, to look at a broad range of solution sets

2.3.2 Explore trade space and margins fully before focusing on a point design and too small margins

2.3.3 Anticipate and plan to resolve as many downstream issues and risks as early as possible to prevent downstream problems

2.3.4 Plan early for consistent robustness and *right the first time* under "normal" circumstances, instead of hero-behavior in later "crisis" situations

2.4 Plan to Develop Only What Needs Developing

2.4.1 Promote reuse and sharing of program assets: Utilize platforms, standards, buses, and modules of knowledge, hardware and software

2.4.2 Insist that a module proposed for use is robust before using it

2.4.3 Remove show-stopping research/unproven technology from critical path, staff with experts, and include in the Risk Mitigation Plan

2.4.4 Defer unproven technology to future technology development efforts or future systems

2.4.5 Maximize opportunities for future upgrades, (e.g., reserve some volume, mass, electric power, computer power, and connector pins), even if the contract calls for only one item

2.5 Plan to Prevent Potential Conflicts with Suppliers

2.5.1 Select suppliers who are technically and culturally compatible

2.5.2 Strive to develop a seamless partnership between suppliers and the product development team

2.5.3 Plan to include and manage the major suppliers as a part of your team

2.5.4 Have the suppliers brief the design team on current and future capabilities during conceptual formation of the project

2.6 Plan Leading Indicators and Metrics to Manage the Program

2.6.1 Use leading indicators to enable action before waste occurs

2.6.2 Focus metrics around customer value, not profits

2.6.3 Use only a few simple and easy-to-understand metrics and share them frequently throughout the enterprise

2.6.4 Use metrics structured to motivate the right behavior

2.6.5 Use only those metrics that meet a stated need or objective

Lean Principle 3: Flow

3.1 Execute the Program According to the INCOSE Handbook Process

3.2 Clarify, Derive, and Prioritize Requirements Early and Often During Execution

> 3.2.1 Since formal written requirements are rarely enough, allow for follow-up verbal clarification of context and need, without allowing *requirements creep*
>
> 3.2.2 Create effective channels for clarification of requirements (possibly involve customer participation in development IPTs)
>
> 3.2.3 Listen for and capture unspoken customer requirements
>
> 3.2.4 Use architectural methods and modeling for system representations (3D integrated CAE toolset, mockups, prototypes, models, simulations, and software design tools) that allow interactions with customers as the best means of drawing out customer requirements
>
> 3.2.5 *Fail early - fail often* through rapid learning techniques (prototyping, tests, digital preassembly, spiral development, models, and simulation)
>
> 3.2.6 To align stakeholders, identify a small number of goals and objectives that articulate what the program is set up to do, how it will do it, and what the success criteria will be, and repeat this process consistently and often

3.3 Front Load Architectural Design and Implementation

> 3.3.1 Explore multiple concepts, architectures, and designs early
>
> 3.3.2 Explore constraints and perform real trades before converging on a point design
>
> 3.3.3 Use a clear architectural description of the agreed-upon solution to plan a coherent program, and engineering and commercial structures
>
> 3.3.4 All other things being equal, select the simplest solution ("Any fool can make anything complex but it takes a genius and courage to create a simple solution"—Albert Einstein.)
>
> 3.3.5 Invite suppliers as trusted program partners to make a serious contribution to SE, design, and development

3.4 Systems Engineers to Accept Responsibility for Coordination of PD Activities

> 3.4.1 Promote maximum seamless teaming of SE and other PD engineers
>
> 3.4.2 Systems Engineers to regard all other engineers as their partners and internal customers, and vice-versa
>
> 3.4.3 Maintain team continuity between phases to maximize experiential learning
>
> 3.4.4 Plan for maximum continuity of Systems Engineering (SE) staff during the program

SAME AS 2.3.4

5.2.2 Promote excellence under "normal" circumstances instead of hero behavior in "crisis" situations

5.2.3 Use and discuss failures as opportunities for learning, emphasizing process and not people problems

5.2.4 Treat any imperfection as an opportunity for immediate improvement and a lesson to be learned, and practice frequent reviews of lessons learned

5.2.5 Maintain a consistent, disciplined approach to engineering

5.2.6 Promote the idea that the system should incorporate continuous improvement in the organizational culture, but also ... (continued in 5.2.7)

5.2.7 ... balance the need for excellence with avoidance of overproduction waste (pursue refinement to the point of assuring value and *right the first time*, and prevent over-processing waste)

5.2.8 Use a balanced matrix/project organizational approach. avoiding extremes, such as territorial functional organizations with isolated technical specialists, and all-powerful IPTs separated from functional expertise and standardization

5.3 Use Lessons Learned from Past Programs for Future Programs

5.3.1 Maximize opportunities to make each next program better than the last

5.3.2 Create mechanisms to capture, communicate, and apply experience-generated learning and checklists

5.3.3 Insist on workforce training in root cause and appropriate corrective action

5.3.4 Identify best practices through benchmarking and professional literature

5.3.5 Share metrics of supplier performance with them so they can improve

5.4 Develop Perfect Communication, Coordination, and Collaboration Policy across People and Processes

5.4.1 Develop a plan and train the entire program team in communications and coordination methods at the program beginning

5.4.2 Include communication competence among the desired skills during hiring

5.4.3 Promote good coordination and communications skills with training and mentoring

5.4.4 Publish instructions for email distributions and electronic communications

5.4.5 Publish instructions for artifact content and data storage—central capture versus local storage, and for paper versus electronic—balancing between excessive bureaucracy and the need for traceability

List of Figures and Numbered Text Boxes

FIGURES

NUMBERED TEXT BOXES

Chapter **1**

Introduction

1.1 INTRODUCING LEAN SYSTEMS ENGINEERING AND LEAN ENABLERS FOR SYSTEMS ENGINEERING

This book was written for two purposes: to popularize the emerging field termed *Lean Systems Engineering (LSE)*, and to serve as a textbook and reference for the first major product created under it: *Lean Enablers for Systems Engineering (LEfSE)*.

The discipline of *Systems Engineering* (SE) was created to help with the development of complex systems that must work unconditionally. It was first shaped in the ballistic missile program by Si Ramo[1] and Dean Wooldridge in 1954, with the first formal contract to perform systems engineering and technical assistance (SETA). Under this contract, Ramo and Wooldridge developed some of the first principles for SE and applied them to the ballistic missile program—considered one of the most successful major technology development efforts ever undertaken by the U.S. government [Jacobsen, 2001]. SE is the practical engineering realization based on *systems thinking*, a comprehensive design process of a system that satisfies all customer needs during an entire system life cycle. The process demonstrates reliable

---

[1] Si Ramo once told F. Brown [SELP Director, Loyola Marymount University, personal communication] the first use of systems engineering in modern times could be said to have started at AT&T when they faced assembling a world-wide telephone system. AT&T, however, did not consider it systems engineering and did not use that name.

Lean for Systems Engineering with Lean Enablers for Systems Engineering,
First Edition. Bohdan W. Oppenheim.
© 2011 John Wiley & Sons, Inc. Published 2011 by John Wiley & Sons, Inc.

execution of system development programs and leads to extraordinary technological successes in space, air, naval, and ground systems and weapons. The International Council on System Engineering (INCOSE), the professional society of Systems Engineers, offers the following definition of Systems Engineering:

> *Systems Engineering (SE)* is an interdisciplinary approach and means to enable the realization of successful systems. It focuses on defining customer needs and required functionality early in the development cycle, documenting requirements, and then proceeding with design synthesis and system validation while considering the complete problem: operations, cost and schedule, performance, training and support, test, manufacturing, and disposal. SE considers both the business and the technical needs of all customers with the goal of providing a quality product that meets the user needs.
>
> —INCOSE

Balance between technical and business success is one of the critical aspects of high-quality SE. Regretfully, during the past decade, in some government programs inadequate incentives tended to throw the balance off. Namely, system or mission assurance was incentivized while short program schedule and low cost were not. In consequence, in many recent programs the schedule and budget were exceeded, and the final schedules and budgets were significantly extended beyond comparable programs of earlier periods. In extreme cases, instead of *engineering complex systems*, the SE process has deteriorated to a *complex bureaucracy of program artifacts* with no technical success in sight. This book describes how *Lean Thinking* is applied as a "rescue" to SE, to achieve both technical and business successes.

Lean Thinking (or, briefly, *Lean*) is a holistic paradigm that originated at Toyota[2] and focused on delivering value to the customer while removing waste from all activities.

---

[2]Throughout its history, Toyota has demonstrated an amazing and unprecedented record of quality and has served as the ideal for best corporate practices in a broad range of activities—product development, manufacturing, supply chain management, enterprise management, and exemplary human relations. The two well-published pillars of Toyota success are *Respect for People and Continuous Improvement*. A good fraction of the enablers presented in his book have been based on these extraordinary Toyota practices. The trust in Toyota was temporarily shaken in late 2000s when media started publishing stories about the unintended acceleration of Toyota cars. In January 2010, Mr. Akio Toyoda, the newly appointed CEO of Toyota Motor Co., during the public hearing by U.S. Congress, started with words of apology for the apparent unintended acceleration problems of some recent Toyota cars. In his testimony he said: "Quite frankly, I fear the pace at which we have grown may have been too quick. I would like to point out here that Toyota's priority has traditionally been the following: *First; Safety, Second; Quality, and Third; Volume*. These priorities became confused ... and I am deeply sorry for any accidents that Toyota drivers have experienced" (ABC News, Feb. 24, 2010). Media accused Toyota of ignoring the problem and blamed the company for hundreds of accidents, including fatal ones. This dramatic story was reversed in 2011. In early 2011 J. Liker and T. Ogden published a book *Toyota Under Fire* which provided evidence completely exonerating Toyota. The book quotes the results of

Lean dramatically reduced car development and production times and costs while increasing product quality and stakeholder satisfaction, and became a standard practice in most manufacturing industries. Inspired by these successes, *Lean* was applied in many other domains, including supply chain management, product development, administration, accounting, healthcare, and government, in each case creating significant well documented benefits.

> *Lean Thinking (Lean):* "Lean Thinking is the dynamic, knowledge-driven, and customer-focused process through which all people in a defined enterprise continuously eliminate waste with the goal of creating value."
>
> —LAI EdNet

Lean Systems Engineering (LSE) is an emerging field representing the synergy of *SE* and *Lean*.

> *Lean Systems Engineering (LSE):* The application of lean wisdom, principles, practices and tools to systems engineering in order to enhance the delivery of value to system's stakeholders.
>
> —LSE Working Group

The *Lean* in Lean SE should be regarded as the process of amending the well-established, traditional SE process with the wisdom of *Lean Thinking*, rather than replacing SE with a new body of knowledge. Put emphatically:

> *Lean Systems Engineering:* does not mean "less systems engineering," but rather more and better SE, with better preparations of the enterprise processes, people and tools; better program planning and frontloading; better

---

a major study performed on request of U.S. Congress by NASA under contract to the U.S. National Highway and Transportation Safety Board. The results were announced by Secretary Ray LaHood on Feb.8, 2010. NASA [2011] results were summarized as follows: "Two mechanical safety defects were identified by NHTSA . . . sticking accelerator pedals and a design flow that enabled accelerator pedals to become trapped by floor mats. These are the only known causes for the reported unintended acceleration incidents. Toyota recalled nearly 8 million vehicles in the U.S. for these two defects. The Liker and Ogden [2011] book indicates that the numerous reported cases of unintended acceleration were caused by drivers using a wrong pedal, a phenomenon known to all car companies and highway police alike; and that there was only one proven deadly accident due to pedal entrapment, and one other accident caused by the car owner using a wrong floor mat. Thus, strictly speaking, Mr. Toyoda's apology turned out to be unnecessary, but, characteristically for Toyota, it was never withdrawn. In addition, Toyota instituted numerous additional organizational safeguards and precautions, including a brake override system that stops the car even when the gas pedal is wide open. This dramatic episode demonstrates the uninterrupted focus on safety, quality, and customer satisfaction by Toyota.

workflow management; and better program management and leadership with higher levels of responsibility, authority, and accountability. The benefits of *LSE* are immediately visible in better SE processes, as well as in the stream-lined execution of the program. There is less waste, waiting, and rework; more predictable and robust program flow; lower overall program cost and shorter schedule, and creation of better value to the customer.

The fundamental feature of *LSE* is to perform all preparations and planning of the people, processes, tools, and individual tasks well enough so that the tasks can be executed *right the first time*,[3] creating customer value while minimizing waste. Under the *LSE* philosophy, system success or mission assurance is non-negotiable, and any task which is legitimately needed for success must be included in the program. It should be well-planned and executed with minimum waste.

Lean Enablers for Systems Engineering (LEfSE) is the first major practical prod-uct created under LSE, and is the main topic of this book. LEfSE is a checklist of 147 practices formulated as dos and don'ts of SE and some closely related aspects of Product Development (PD) and Enterprise Management (EM), including supply chain management. The intent of LEfSE is to offer comprehensive and actionable practices to the profession, with the objective of improving overall performance of SE and the PD program it serves; strengthen the value created by the program; increase the level of satisfaction of stakeholders; and reduce waste, program cost, and time. Most of the effort in developing LSE and the LEfSE was conducted by the LSE Working Group (LSE WG) of INCOSE. This was an intensive and rigorous two-year process by 14 experts, supported by members of the WG, and endorsed by surveys and by benchmarking with recent NASA and Government Accountability Office (GAO) recommendations for system development.

The surveys confirmed that practitioners regard LEfSE as both important, and not yet widely used. Benchmarking showed excellent convergence between LEfSE on one hand and NASA and GAO recommendations on the other, with LEfSE usually more detailed, comprehensive, and actionable than the GAO recommendations. The development of LEfSE is summarized in Chapter 6 of this book.

Version 1.0 of LEfSE was released to INCOSE and to the public in January 2009. In 2010 the LEfSE were recognized with two following prestigious awards:

- The INCOSE Product of the Year
- The Shingo Award for Research and Publication

These independent recognitions from both *SE* and *Lean* communities are regarded as an important step in validating LEfSE.

---

[3]The expression *right the first time* refers to both single-pass tasks as well as engineering iterations and other complex activities, which also must be regarded as tasks that need to be well planned, designed, and executed so that they can be completed robustly in a predictable time without wasteful rework.

After the LEfSE were released to the public on the INCOSE web page,[4] the author and several of his colleagues on the development team were invited to offer numerous (forty at the time of this writing) workshops, tutorials, webinars, and seminars to industry, governments, INCOSE meetings, and universities in the U.S., China, Finland, France, Italy, Israel, Netherlands, Norway, Poland, Singapore, Sweden, and the United Kingdom. Over 2000 participants took part in these professional events, generally excellently received. In many of these sessions, participants asked for more explicit information about the enablers than was possible to present in the short sessions. This book is a direct response to these requests.

As already mentioned, this book is intended as both an introductory textbook on LSE and a textbook and reference for Lean Enablers for SE. The material presented is broad in scope and it applies to the development of all types of complex systems in all domains, both in commercial and government programs. We hope that LEfSE will benefit Systems Engineers and other PD engineers and managers, professionals involved in systems acquisition both in industry and government, project managers, and faculty and students of these fields.

Even though the material presented in this book applies to all domains and all program types, government programs are mentioned most often for two reasons. First, SE is mandatory in such programs while it is not in commercial programs, thus the body of experience from government programs is significantly larger. Second, the amount of waste in governmental programs tends to be significantly higher than in commercial programs, which makes it a fertile ground for Lean thinkers.

The author chooses to explain selected enablers by comparing them to some current industrial and governmental practices that are less than perfect. The author hopes that the reader will accept this approach as constructive criticism leading to improvement of our programs, rather than "bashing" of the traditional practices. To repeat with emphasis: The traditional SE process is regarded as sound, capable of delivering successful complex systems, but not as good as it could be, and which we therefore propose to improve using Lean.

Both SE and Lean are a mix of art and science, and the art of effective teaming plays a critical role in SE. Some programs can be compared to the effort of climbing Mt. Everest or competitive racing in a storm—extraordinary teams pursuing difficult challenges against heavy odds. The Apollo, U2, F117, Boeing 747, nuclear submarine, and the Prius car programs come to mind as examples. Such programs inspire young people to choose a career in engineering. In sad contrast, creation of bureaucratic SE artifacts in isolated cubicles is not what inspiriting engineering was promised to be. The best programs must involve leadership, teamwork, enthusiasm, passion, inner energy, and joy in addition to hard competence and experience. The author tried to include these intangible elements in the book, in order to enthuse the reader about the potential of Lean,

---

[4]The web page of the Lean Systems Engineering Working Group of INCOSE is public and accessible to everyone. Appendix 1 provides the web address and a summary of the site content.

which is capable of drawing the best from us and motivating us to engineer extraordinary systems using extraordinarily efficient programs.

SE is a part (in fact, the most critical part for success) of the larger Product Development (PD) effort. SE has been called the nervous system of PD [Hitchens, 2007], planning, controlling, and monitoring in real time all functions of the PD "body." However, many enterprises have structured SE to be a separate function (department) that supports all their PD programs. This led to a linguistic dichotomy, "SE and PD". In this book the term "SE and PD" should be interpreted to mean "SE and other elements, parts, activities, etc. of PD."

The traditional SE process is described in a number of manuals: INCOSE SE Handbook, the ISO 15288 standard, NASA and Department of Defense SE Manuals, Defense Acquisition University manuals, and numerous manuals created by individual defense and civilian companies. Arguably, the manuals describe essentially the same body of knowledge and the same *traditional* SE process with varying degrees of detail, emphasis, and user friendliness. Since the present project has been carried out under the auspices of INCOSE, and because the author regards the INCOSE SE Handbook v.3.2, 2010 as an excellent and user-friendly document, the present text makes references to that document only. Version 3.2 of the Handbook includes a short chapter on Lean Thinking. The first enabler of LEfSE explains that practitioners should continue using all processes described in the Handbook, adding Lean practices listed in LEfSE.

1.2   ORGANIZATION OF THE BOOK

The book is organized as follows.

Chapter 2 takes the reader through a brief historical review of earlier industrial paradigms, setting Lean Thinking in the context of Total Quality Management (TQM), Concurrent Engineering (CE), and Six Sigma.

Chapter 3 presents Lean Thinking fundamentals, including the concepts of value, waste, and the process of creating value without waste, as captured into the six Principles of Lean Thinking. The Principles are called Value, Map the Value Stream, Flow, Pull, Perfection, and Respect for People. The fundamentals are included to make the book self-contained, but the reader unfamiliar with Lean will surely benefit from reading the transformational book *Lean Thinking* by J. Womack et al. [1996]. Readers who understand Lean manufacturing but have not been exposed to Lean Product Development (LPD) should read this chapter because it sets Lean fundamentals in the PD context.

The ability to recognize waste in Product Development is a critical skill in LSE. Waste should be regarded as a productivity reserve. Exposing waste aids in streamlining the program and creating time and cost buffers, thus helping the program to meet schedule and budget. Some wastes are self-evident and incontestable (e.g., waiting, or defects), but others may be less obvious, particularly to Systems Engineers who have to struggle for enough budget and time to execute a program well. Therefore, the discussion of waste is an important part of the book.

Chapter 4 describes the fast growing field of Lean Product Development, including a review of literature and the progress to date. Arguments are presented in support of the thesis that the application of Lean Thinking to Product Development has become mature enough for immediate use in programs.

In Chapter 4, Section 4.2 describes a highly efficient holistic process called Lean Product Development Flow (LPDF) developed by Oppenheim [2004] for organizing smaller and low-risk PD programs. LPDF was created as a contribution to LSE, but is an earlier and totally separate effort from LEfSE.

Chapter 5 presents the emerging field of LSE. We start with a review of recent literature documenting both successes and problems in technology programs and justify the need for better SE, with less waste, leading to LSE. We also introduce the INCOSE Lean SE Working Group. This chapter ends with a discussion of value in LSE.

Chapter 6 describes the strategy used for the development of LEfSE, and the endorsement steps. The interested reader will find details of the development process in a comprehensive journal article by Oppenheim, Murman, and Secor (2010), which are not repeated here. We only briefly describe the endorsement efforts by peers, including a survey completed by practicing Systems Engineers, and benchmarking of LEfSE with recent recommendations, which were published by NASA and Government Accountability Office (GAO) when this project was nearing completion. The reader will find that the survey ranked all enablers as (paraphrasing) either *important* or *very important*, and not yet used widely, confirming the importance and the need for LEfSE in the SE practice. The LEfSE have been found to be totally convergent with the NASA and GAO recommendations, but are more comprehensive, detailed, and actionable.

The version of LEfSE presented in this book is regarded as mature enough for presentation to the professional community. However, the intent is to continue involving the community of practice in gathering data and experiences to continue improving the LEfSE product.

Chapter 7 is the main and longest part of this book. It presents the enablers in a standard tabular form for easy reference. The enablers are organized under the six Lean Principles. The text under each Lean Principle begins with a short summary of the listed enablers.

Each enabler or closely related group of enablers is described in a separate table listing the value promoted and the waste prevented by the enabler(s), an explanation of the enabler(s), recommended implementation, lagging factors, and recommended reading list.

In most cases, each table covers only one enabler. In other tables, several closely related enablers are discussed together in order to facilitate reading and implementation.

Chapter 8 contains general guidance for implementing LEfSE. We hope by that time, the reader will be familiar with LEfSE in Chapter 7. We suggest how to select and prioritize the enablers for implementation, and recommend practical steps.

In Section 8.2 we include some early feedback and results from the companies and programs trying to implement LEfSE, and other relevant studies. At the time of this writing, some of these success stories are still fragmentary and lacking scholarly rigor, but they demonstrate the powerful potential of the Lean approach. The sources claim that with only a few Lean enablers implemented, various program time/cost elements were reduced between 20 to 40%, achieved a ROI of 5, and improved the workplace morale, product quality, and customer satisfaction.

Appendix 1 summarizes the content of the INCOSE Lean SE Working Group webpage.

Appendix 2 presents a mapping of Lean enablers onto the 26 processes listed in the INCOSE SE Handbook version 3.2. This mapping project was initiated by INCOSE, but has not yet been approved by that organization. This appendix also contains a short discussion of program lifecycle frameworks other than the SE processes.

Two glossaries are included—one for abbreviations, and one for idioms, colloquialisms, and foreign expressions used in the fields of SE and Lean. These items are **written in bold italic font** in the text. *The italic font alone is used when introducing important new terms, and for emphasis*. **Bold font** (as in the next sentence) is used for greater visibility when navigating through these pages.

How to use this book? Those readers who are eager to become productive can skip Chapter 2, with its historical review, and Chapter 6, which covers the development of LEfSE. Readers who are experts in Lean Thinking and familiar with fundamentals of Lean Product Development may go directly to Section 4.2 of Chapter 4 for use as a reference for the LPDF method, and to Chapter 7, as a reference for improving the practice of SE with LEfSE. Readers who are new to Lean Thinking should first read Chapters 1 and 3–5.

Chapter **2**

A Brief History of Recent Management Paradigms

"A dwarf standing on the shoulders of a giant may see farther than a giant himself".
—Didacus Stella

2.1 FROM TQM TO SIX SIGMA AND LEAN

Three 1970s era events in the United States provided a fertile ground for subsequent dynamic changes of industrial paradigms: (1) oil embargoes, which made small cars attractive to consumers; (2) the overtaking of the consumer electronics and auto markets by higher quality and less expensive Japanese imports; and (3) a widespread perception that U.S. manufacturing was falling behind international competition, particularly in quality. In 1980, NBC TV broadcast a two-hour program titled "If Japan Can... Why Can't We?" opening U.S. eyes to a new management paradigm named *"Total Quality Management"* (TQM), which was sweeping industry by storm in the early 1980s [NBC, June 24, 1980]. Led by Edward Deming [1982] this was an attempt to adopt successful Japanese industrial management methods for U.S. industry. A strong message of TQM was that pursuit of higher quality is compatible with lower costs. Inexpensive and high-quality automobiles and consumer electronic goods imported from Japan made this notion self-evident to U.S. consumers, but not necessarily to U.S. industry.

Lean for Systems Engineering with Lean Enablers for Systems Engineering,
First Edition. Bohdan W. Oppenheim.
© 2011 John Wiley & Sons, Inc. Published 2011 by John Wiley & Sons, Inc.

TQM emphasized a *total* approach to quality by integrating management, processes, and tools, and developing a business strategy focused on customer satisfaction. It promoted continuous improvement of all processes and popularized improvement tools such as a bottom-up employee suggestion system, quality circles, quick reaction *Kaizen* teams, and process variability reduction using Statistical Process Control and Design of Experiments. TQM emphasized the importance of corporate culture based on respect for people, and employee empowerment as prerequisites for continuous improvement and relied heavily on self-motivation of employees. It also promoted designing quality into both products and processes, rather than relying on the final inspection.

TQM received strong support from the U.S. federal government, including the Department of Defense [DoD, 1988]. Following the Japanese E. Deming Award, the Department of Commerce initiated the Malcolm Baldrige Award in 1987 as a motivational recognition of the best U.S. companies in three categories: manufacturing, service, and small business. The award uses a point score that was based on the above TQM elements. About eight years into the TQM period, Costello [1988], in his capacity as Under Secretary of Defense for Acquisitions, presented an alarming report to the Secretary of Defense about serious problems plaguing the U.S. commercial and military industrial base, including foreign competition, poor quality of both products and business management, fragmented research and development, low quality of public education, and declining numbers of engineers and scientists. Four years later, Costello [1992], reinforced this report with a white paper about the state of U.S. industry, recommending wide-ranging improvement of the manufacturing sector, streamlining regulations, better means for technology sharing, and aggressive support for small business.

In 2000, the International Standards Organization (ISO) issued a quality standard denoted ISO 9000:2000, which captures many of TQM elements.

The application of TQM to U.S industry had mixed outcomes. While quality improved, especially in the auto industry, profits did not follow proportionately. Even the quality improvements alone failed in many companies that tried TQM [Paton, 1994]. The earlier pessimism in the manufacturing industry continued and contributed to the subsequent export of nearly 60% of commercial manufacturing to cheaper labor countries.

The lack of widespread business success made TQM vulnerable to criticisms and opened the way to new ideas. *Business Week* [Byrne, 1997] declared TQM a "dead fad," blaming TQM's "lack of teeth" in implementation. While today the term TQM has receded, most of the key elements of TQM have endured and are integral to Lean Thinking [Murman et al., 2002].

In late 1980s the *Concurrent Engineering* (CE) industrial paradigm became popular and was proposed as a way to shorten the weapons system acquisition cycle [Winner et al., 1988]. CE promoted simultaneous and integrated design of product and subsequent phases (manufacturing, assembly, operations, etc.), replacing the traditional disjointed and serial effort. An important component of CE was multifunctional design teams, sometimes called *Integrated Product Teams* or IPTs, which included representatives from subsequent phases in the upfront engineering design. CE, when effectively implemented with electronic design tools and workforce training, led to dramatic reduction in design rework and, consequently, cost and schedule [see Hernandez, 1995]. CE contributed major improvements to U.S. product development and engineering, and spawned significant new design methodologies [see Clausing, 1994; Ulrich and Eppinger, 2008]. TQM and CE made important contributions to major new aircraft products in the 1990s, such as the Boeing 777 and Cessna Citation-X [Haggerty and Murman, 2006]. As with TQM, CE principles are embedded in current day Lean Thinking [Murman et al., 2002; McManus 2004].

In the early 1990s, TQM evolved into another quality initiative called *Six Sigma*, arguably with "better teeth." According to Wedgewood [2007], "Six Sigma is a systematic methodology to home in on key factors that drive performance of a process, set them at the best levels, and hold them there for all time." Originating at Motorola and relying on rigorous measurement and control, Six Sigma focused on systematic reduction of process variability from all sources of variation: machines, methods, materials, measurements, people, and environment [Murman et al., 2002]. Like TQM, Six Sigma aims to achieve predictable, repeatable, and capable processes, and defect-free production, where parts and components are built to exacting specifications. But unlike the motivational TQM, it achieves this by rigorous data collection and statistical analysis, as well as rigorous training of leaders.[1] Following a similar path as TQM and CE, the Six Sigma movement spawned new Design for Six Sigma methodologies (e.g., Yang and El-Haik, 2003).

Six Sigma was not free of problems: it often was implemented with a costly bureaucracy, introducing *the waste of measuring waste*, and was criticized for being too top-down and for displacing two other critically important continuous improvement tools of TQM—small quick-reaction **Kaizen** and the bottom-up employee suggestion system, which Toyota credits for half of its success [Oppenheim, 2006]. Six Sigma can also be prone to sub-optimization by focusing too narrowly on process improvement for a process that may not be needed. Murman described this deficiency as "a focus on *the job being done right*, but not necessarily on *the right job*" [Murman et al., 2002]. It was the next step in the industrial evolution, called Lean, which provided the integrated focus on *the right job* and *doing the job right*, and also on the management culture needed for both.

The term *Lean* as an industrial paradigm was introduced in the United States in the bestselling book *The Machine That Changed the World: The Story of Lean Production*, published by the MIT International Motor Vehicle Program [Womack et al., 1990] and elegantly popularized in their second bestseller *Lean Thinking: Banish Waste and Create Wealth in Your Corporation* [Womack and Jones, 1996]. The authors identified a fundamentally new industrial paradigm based on the Toyota Production System. The paradigm is based on relentless elimination of waste from all enterprise operations and requires the continuous improvement

---

[1]Following the *ju-jitsu* language, Six Sigma leaders are designated by "belts" of various colors denoting different levels of training and experience.

cycle that turns all front-line workers into problem solvers to eliminate waste. Lean strives for minimum waste to deliver high quality and defect-free products meeting customer demand just-in-time, at the rate ordered, with minimum inventories, at minimum cost, and in minimum time. Lean is driven by a unique management culture of respect, empowerment, openness, and teamwork. Factories adopting Lean observed direct and dramatic improvements of operations and increases in profits. Womack and Jones [1996] described six manufacturing case studies that demonstrated reductions of cost, lead time, and inventory of up to 90%, with simultaneous improvements in product quality and work morale across a wide range of company types and sizes. More dramatically, lead time and cost reductions on the order of 30 to 50% were realized routinely after only a few days of implementation on the factory floor, by a simple rearrangement of machines into the flow [LEI, 2007]. After the multi-year implementation efforts of TQM, CE, or Six Sigma, this was a revelation. Within a few years, Lean production has become the established manufacturing paradigm pursued by all competitive factories.

Lean Thinking, or more briefly Lean, as used in this book, is an evolutionary industrial paradigm incorporating elements from earlier paradigms of TQM and CE, as well as elements of Six Sigma. In common with TQM and CE, Lean focuses on designed-in/built-in quality, Edward Deming continuous improvement cycles, and engagement of frontline workforce in process improvement. It goes beyond TQM and CE to adopt a value stream focus, connecting tasks and processes into the flow of value-adding effort and a relentless pursuit of waste elimination. While Lean, TQM, and CE all focus on process improvement, Lean particularly focuses on streamlining flow between the processes. Sharing with Six Sigma a data-driven approach to eliminate process variation, it differs by being more bottom-up in its improvement strategy and less reliant on formalized qualifications of improvement experts. As with the other improvement paradigms, successful Lean implementation relies on committed leadership and an enterprise-wide approach across all functions, including systems engineering.

As already mentioned, Lean incorporates many of the TQM, CE, and Six Sigma principles and practices. However it goes beyond them to adopt a holistic value stream approach and relentless waste elimination. The value stream represents the linked end-to-end activities that turn raw material or information into products, systems, and services needed by the customer. Waste represents those activities that do not directly contribute to customer value. Often these are activities taking place between *valued added* (VA) activities, such as waiting. Lean strives for optimum flow with no blockages or unplanned rework.[2]

Toyota's original "father" of Lean, [Ohno, 1988, foreword by N. Bodek, page ix] summarized it thus: "All we are doing is looking at the time line from the

---

[2]Some might ask how the focus on flow differs from Henry Ford's moving line mass production or from "rhapsodized industrial engineering." Indeed there are common elements. However there are important distinctions. Lean emphasizes the importance of the frontline workers as problem solvers, unlocking the enormous human resource potential for process improvement. Lean also focuses on **single piece flow** with minimum inventories, which leads to cellular work arrangements. This is contrasted to the method of **batch and queue** practiced in traditional production.

moment the customer gives us an order to the point when we collect the cash. And we are reducing that time line by removing the non-value added wastes."

The successes of Lean in repeatable manufacturing led to a popular misconception that Lean does not apply to *one-off* applications such as engineering projects or SE. In such projects, the deliverables and work content of engineering tasks are indeed *one-off* but many processes and individual tasks should use repeatable logic based on best engineering practices. For example, the established process for a modal analysis of a structure involves the same steps: define geometry, boundary conditions, and material properties; perform finite-element meshing; calculate the modal eigenvectors and eigenvalues; identify dangerous modes; and write the report. Such established processes exist for the vast majority of engineering applications, including SE. Therefore, it makes sense to apply the basic principles of Lean Thinking to SE.

Lean Thinking has been applied to such diverse work environments as manufacturing [e.g., Womack and Jones, 1996], product development [e.g., Ward, 2007], engineering [e.g., McManus et al., 2007], supply chain management [e.g., Bozdogan, 2004], healthcare [e.g., Graban, 2008], education [e.g., Emiliani, 2004], administration [e.g., Carter, 2008], and Enterprise Management [e.g., Jones, 2006; LESAT, 2001]. This book describes the application of Lean to Systems Engineering and how that relates to Lean Product Development.

2.2 LEAN SIX SIGMA

Lean and Six Sigma both appeared in the post-TQM mid-1990s as seemingly competing process improvement approaches. Six Sigma, identified with Motorola and subsequently with GE, gained investor visibility and popularity. Lean, identified with Toyota, was incorrectly looked upon as limited to high-volume manufacturing applications. While Six Sigma focuses on a disciplined, top-down approach to eliminating all forms of variation, Lean focuses on value streams and relentless elimination of waste through optimizing flow. The latter relies on the former to eliminate impediments to flow, and in fact the basic principles of the two approaches are synergistic. By early 2000, most organizations adopted a blended version of the two bodies of knowledge and crafted them to meet their particular needs. Names such as Lean Six Sigma, Lean Sigma, and other less obvious combinations appeared. Today, most organizations have harmonized Lean and Six Sigma. In this book, we will continue to use the Lean nomenclature, not to exclude Six Sigma thinking, but to treat the new field *Lean Systems Engineering* and its first major product *Lean Enablers for Systems Engineering* as an integrated body of knowledge based on Lean, Six Sigma, high-performance work systems, and other process improvement approaches.

Chapter 3

Lean Fundamentals

"All we are doing is looking at the time line from the moment the customer gives us an order to the point when we collect the cash. And we are reducing that time line by removing the non-value added wastes."

—Taiichi Ohno, Toyota's original "father" of Lean

Three concepts are fundamental to the understanding of Lean Thinking: **value, waste**, and the process of creating value without waste, captured into the so-called six **Lean Principles**. The concepts are described in this chapter in the general context of product development. The concepts are explained in a self-contained style to free the reader from the need to refer to other sources, but the reader who is new to Lean would benefit from first reading the classic book *Lean Thinking* [Womack and Jones, 1996].

A formal definition of Lean has been presented in Chapter 1. However, when considering the present Lean fundamentals, it is useful to think of Lean simply as "creation of best value with minimum waste."

3.1 VALUE

At its simplest, *value* is what the customer says it is, considers important, and is willing to pay for. In simple applications, the customer states what he or she

Lean for Systems Engineering with Lean Enablers for Systems Engineering,
First Edition. Bohdan W. Oppenheim.
© 2011 John Wiley & Sons, Inc. Published 2011 by John Wiley & Sons, Inc.

needs, and the contractor makes it and delivers it, hopefully satisfying or even delighting the customer. This works well when buying ice cream, but is vastly more challenging when ordering a new, complex technological system, especially for military users.[1]

> *Value* is what the customer says it is, considers important, and is willing to pay for.

In government programs, the number of stakeholders often reaches thousands of individuals in numerous communities of users and acquisition stakeholders; the prime contractor and suppliers in the entire value chain; as well as other stakeholders—politicians, lobbyists, shareholders, banks, and so on. Different stakeholders promote different aspects of value dear to them, often in conflict with one another. These factors make the value capture and contract formulation a significant challenge and a costly process. Yet, it must be done perfectly, or the subsequent program will suffer delays, added costs, frustrations, and, in extreme cases, program closure or failure. It is critical that everybody involved in the process be focused on capturing the final value proposition with the absolute best of competence, wisdom, experience, and consensus. A value definition must be crystal clear, unambiguous, complete, representing customer need during a system lifecycle, and allowing effective channels for value clarification without causing *requirements creep*.

In Section 8.2 we quote from a study by Honour [2010] that states that even properly scoped and funded SE activities have little effect on the technical quality of the product system. Apparently, the reason is they aim for minimum requirements compliance rather than technical optimization. This profound finding is interpreted here as a call for a much better definition and subsequent optimization of value than it is practiced in recent programs. This book contains ideas as to how that can be accomplished without raising the program cost or lengthening the schedule.

3.2   WASTE

The ability to identify and eliminate waste it is a critical skill for Lean.

---

[1] Anyone who has been involved in even a minor house remodeling knows that the formulation of value that leads to the *right the first time* execution within budget and schedule is difficult; cynics say impossible. Misunderstandings are ubiquitous and usually end up in a fee and schedule increase, even though only two parties are involved (owner and contractor), the job description is simple, and the technology mature. Who is at fault for the changes from the baseline? Most likely both parties: The owner's specifications lack crystal clarity and completeness, and the contractor fails (perhaps intentionally) to ask clarifying questions to ensure interpretation is the same as an owner's.

> *Waste* is anything other than the minimum amount of resources (employee time, space, materials, equipment) that are absolutely essential to create value for the customer.
>
> —attributed to Shinichi Suzuki in [Herman, 1981]

Lean Thinking classifies all work activities into three categories [Womack and Jones, 1996; LAI Lean Academy, 2008]:

1. **Value Added (VA)** activities, which must satisfy the following three conditions:
   - Transform information or material, or reduce uncertainty (cannot be just an unnecessary bureaucratic task that creates no value).
   - The customer must be willing to pay for it (explicitly, or, in more complex programs, implicitly, that is *if the customer understood the details, he or she would approve of this activity*).
   - It is done *right the first time*[2].
2. **Required (also called Necessary) Non-Value Added (RNVA)** activities, which do not meet the above definition, but which cannot be eliminated because they are required by law, contract, company mandate, current technology, or other such reason.
3. **Non-Value Added (NVA)** activities, which consume resources and create no value. They are pure waste (e.g., unneeded reports and emails, idle time, defects that require rework, etc.)

Taiichi Ohno classified waste in manufacturing into seven categories. Several authors have adapted Ohno's seven production wastes for Product Development, including Morgan and Liker [2006] and McManus [2004]. Box 3.1 lists the interpretation by the author in the SE context.

The examples listed in Box 3.1 are intended only to "whet the appetite" of the reader. Any experienced systems engineer should have no trouble expanding this list based on personal experience.

A number of Lean researchers also use additional categories: excessive complexity (which we classify as Processing waste), destructive infighting (which we

---

[2]Note: Engineering iterations, complex studies, and other such reports are included in this statement. When we refer to *right the first time* we mean a task that could have been performed right on the first attempt with proper preparations, planning, coordination, and communication, but these reasonable steps were not performed, so now the task failed and must be repeated, causing rework. Therefore, that first failed attempt must be regarded as waste.

break down into the seven categories), waste of human talent (which we interpret as excessive waiting and overproduction), and so on.

3.3 LEAN PRINCIPLES

The process of creating value without waste has been captured into six Lean Principles titled: **Value, Map the Value Stream, Flow, Pull, Perfection**, and **Respect for People**. The first five were formulated by Womack and Jones [1996]. The sixth is often called "the second pillar of Lean" [Sugimori et al., 1977]. It plays a critically important role for Lean. In this book, it is elevated to the explicit sixth Principle to assure its maximum visibility and profound consideration.

A frequent concern of Systems and other PD engineers unfamiliar with Lean is that Lean applies only to repeated production and does not apply to *one-off* PD products. This impression can easily be dismissed by noticing that each Principle applies to any number of items being created: from a *one-off* engineering effort to high-volume repeated production, placing no limitations whatsoever on the type of process, e.g., production or engineering. As already mentioned in Chapter 1, Lean has been applied in a broad range of work environments.

3.3.1 Principle 1: Value

Capture the 'value' defined by the Customer, who may be either external or internal. The external customer who pays for the system or service defines the final value for the deliverable. Internal customers receive the output of a task or activity and usually don't explicitly pay for it. Either way, the customer is the one who defines what constitutes value. One cannot overemphasize the importance of capturing task or program value with precision, clarity, and completeness before resource expenditures ramp up to avoid unnecessary rework. For programs as lengthy as those in government-complex technology acquisition programs, external factors can change, and customer value expectations may need to be revisited, updated, or revised [Murman et al., 2002]. Clearly, a careful balance is needed. On the one hand, constant change and instability must be avoided or the system costs will grow and schedule will lengthen (e.g., *Space Based Infrared System (SIBRS)* program [GAO, 2007]). On the other hand, customer value expectations or threats may change and an original value proposition could become obsolete (e.g., recent cancellation of further F-22 aircraft production). This is the strongest argument for shorter program schedules.

3.3.2 Principle 2: Map the Value Stream

'Map the value stream (plan the program)' and eliminate waste: Map all end-to-end linked tasks, control/decision nodes, and the interconnecting flows necessary to

Box 3.1 Seven Categories of Waste, Based on Morgan and Liker [2006]

SEVEN WASTES	WHAT IS IT?	PD EXAMPLES
Overproducing	• Producing more than the next process needs • *Reinventing the wheel*	• Creating too much information • Engineering beyond the precision needed • Overdissemination = sending information to too many people (e.g., excessive email distribution) • Sending a volume when a single number was requested • Ignoring expertise
Waiting	• Waiting for information or decisions • Information/decision waiting for people	• Long approval sequences • Waiting for data, test result, information, decision . . . • Late delivery • Poor planning, scheduling, precedence • Unnecessarily serial effort
Conveyance	• Moving information from place to place	• Hand-offs/excessive information distribution • Disjointed facilities, political *made in 50 states*, lack of co-location • Uncoordinated complex document taking so much time to create that it is obsolete when finished
Processing	• Doing unnecessary processing on a task or an unnecessary task	• Stop-and-go tasks • Redundant tasks, reinvention, process variation—lack of standardization • Creating documents that nobody requested • Point design used too early, causing massive iterations

Category	Description	Examples
		• Uncontrolled iterations (too many tasks iterated)
		• Work on a wrong release (information churning)
		• Data conversions
		• Answering wrong questions
		• Many of contractual obligations (e.g., 2D drawings)
		• Unclear or unstable requirements
		• Excessively complex software *monuments* (using complex software when a spreadsheet would do)
Inventory	• A buildup of information that is not being used	• Batching
		• System overutilization
		• Arrival variation
		• Poor configuration management and complicated retrieval
		• Poor 5Ss in office or databases
		• Lacking central release
People motion	• Excessive motion or activity during task execution	• Long travel distances
		• Redundant meetings
		• Superficial reviews
		• People having to move to gain or access information
		• Manual intervention to compensate for the lack of process
Correction	• Inspection to catch quality problems • Fixing an error already made	• The killer "re's": Rework, Rewrite, Redo, Reprogram, Recertify, Recalibrate, Retest, Reschedule, Recheck, Re-inspect, Return, Re-measure . . .
		• Incomplete, ambiguous, or inaccurate information
		• External quality enforcement

realize customer value. During the mapping process, eliminate all non-value added activities, minimize all necessary non-valued activities, and enable the remaining activities to flow without rework, backflow, or stopping (the flow is described under Principle 3). A key concept to grasp in moving from the manufacturing to the engineering domain is that in the former, material is being transformed and moved, while in the latter information is being transformed and moved. The term *information flow* refers to the packets of information (knowledge) created by different tasks and flowing to other tasks (design, analysis, test, review, decision, or integration) for subsequent value adding. There are a number of implications in applying Lean thinking principles, techniques, and tools to a medium that is as fluid as information. Careful detailed planning and program front loading, common or interoperable databases, rapid and pervasive communication of decisions via Intranets or personal communication and frequent **integrative events** for efficient real-time resolution of issues and decision-making, stand-up meetings, or virtual reality reviews are some techniques to keep information flowing. Each task adds value if it increases the level of useful information and reduces risk in the context of delivering customer value. McManus [2004] offers a practical guide for PD value-stream mapping.

The generic term *planning* includes two distinct phases—enterprise preparations and program planning. Corporate enterprises should prepare resources (people, processes, and tools) that will serve all programs. Resources should include an infrastructure for continued employee education and training; creation of the communities of practice; central databases with former design and program data, lessons learned, and knowledge shared; standardization of processes; preparation of the program infrastructure, equipment, and tools; rotation of key people; strategic decisions for subsystem reuse in future programs; and training of employees in the best communication and coordination practices. These activities will serve all programs and should be handled at the corporate level, enhancing the long-term competitiveness of the enterprise. In contrast, the term *PD planning* refers to the planning effort for a specific PD program.

Section 4.2 contains a three-step procedure for mapping simpler and low-risk PD programs or individual program milestones, using maximum concurrency of tasks and a regular frequent cadence of reviews. The procedure is a part of the *Lean PD Flow* (LPDF) process [Oppenheim, 2004].

3.3.3   Principle 3: Flow

'Flow' the work through planned and streamlined value-adding steps and processes, without stopping or idle time, unplanned rework, or backflow. To optimize flow, one should plan for maximum concurrency of tasks, up to near-capacity of an enterprise. Robust capture of value, good enterprise-level preparations, and good program planning are among some necessary conditions for subsequent Lean execution of a program. Detailed planning of a complex program is critical for

Lean, yet it is difficult. It took Toyota several decades to perfect its system, and Toyota employees, who tend to be significantly ahead of others, routinely claim that they are far from perfect.

Legitimate engineering iterations are frequently needed in PD to address *chicken versus egg* technical problems, but they tend to be time consuming and expensive if they extend across disciplines. Lean PD encourages efficient methodology of *fail early – fail often* through rapid architecting and discovery techniques during early design phases. Lean flow also makes every effort to use techniques that obviate lengthy iterations, such as design frontloading, trade space explorations, set designs, modular designs, legacy knowledge, and large margins. Where detailed cross-functional iterations are indeed necessary, Lean flow optimizes iteration loops for overall value, limiting the tasks within loops to only those that experience changes of state, and optimizing their execution for best value [Warmkessel, 2002].

3.3.4 Principle 4: Pull

Let customers 'pull' value. In manufacturing, the ideal pull principle is implemented as the Just-in-Time (JIT) delivery of parts and materials to the needing station and to the external customer. In PD applications, the pull principle has two important meanings: (1) the inclusion of any task in a program must be justified by a specific need from an internal or external customer, and coordinated with them; and (2) the task should be completed when the customer needs the output: excessively early completion leads to "shelf life obsolescence," including possible loss of human memory or changed requirements, and late completion leads to schedule slip. These imply that every task *owner* or engineer be in close communication with his or her internal customers to fully understand their needs and expectations and to coordinate work, modalities, and deliverables. For products complex enough to require systems engineering, one needs both a Lean-thinking customer as well as a Lean-thinking creator. A customer who makes arbitrary demands prevents a Lean outcome, and uncontrolled pull tends to create chaos.

3.3.5 Principle 5: Perfection

Pursue 'perfection' of all processes. Global competition is a brutal "race without a finish line" [Schmidt et al., 1992], requiring continuous improvements of processes and products. Yet, no organization can afford to spend resources improving everything all the time. To clarify the issue, we need to make a distinction between processes and process outputs. Perfecting and refining the work *output* in a given task must be bounded by the overall value proposition (system or mission success and program budget and schedule), which defines when an output is good enough. Otherwise, notorious PD waste of over-processing may occur. Judgments should be made by experienced domain engineers in close coordination with systems engineers who are responsible for overall flow of value. In contrast, engineering and other processes must be continuously improved for never-ending competitive reasons. It is important for the enterprise to understand the distinction between process and product perfection and provide resources accordingly. The next question is which processes to improve. All of them at the same time? Or should some have a higher priority than others? The answer is suggested by two features of Lean: (1) making all imperfections in the workplace visible to all and (2) prioritizing to eliminate the biggest impediments to flow. Seeing problems as they appear in real time is conducive to making better decisions on corrective actions and better prioritization of improvements. When noticed early, imperfections tend to be easier and less expensive to fix; unnoticed early they tend to grow to crisis proportions and require heroic actions to mitigate. Making imperfections visible motivates all to apply continuous improvement in real time [Morgan and Liker, 2006]. The enterprise should create an effective infrastructure for capturing knowledge and lessons learned, and promoting continuous education to make each program better than the last.

3.3.6   Principle 6: Respect for People

Respect for people. A Lean enterprise is an organization that recognizes that its people are the most important resource and is one that adopts high performance work practices [e.g., Gittell, 2003]. In a Lean enterprise, people are not afraid to identify problems and imperfections honestly and openly in real time, brainstorm about root causes and corrective actions without fear, and plan effective solutions together by consensus to prevent the problem from occurring again. When issues arise, the system is blamed and not the messengers. Experienced and knowledgeable leaders lead and mentor, but also empower frontline employees to solve problems immediately. Such an environment requires a culture of mutual respect and trust, open and honest communication, and synergistic and cooperating relationships of stakeholders [Sugimori et al., 1977].

3.4   THE LEAN SYMPHONY OF THE PRINCIPLES

The six Lean Principles fit together like experienced symphony orchestra musicians—in perfect alignment to create the best value with minimum waste, satisfying the program stakeholders. All Principles are critical to success, and no Principle has highest priority. However, these Principles differ among themselves in degree of implementation difficulty. Experienced managers will readily agree that most engineering and technical challenges are easier to address than human relations and corporate culture.

The 'Value Principle' is difficult to formulate in complex PD environments with sufficient clarity and completeness because an understanding of a needed system is often vague and incomplete, yet it must be done right. Success requires best practices of Systems Engineering, a strong cooperation of all stakeholders formulating a value proposition, a robust yet efficient process, and excellent leadership.

The 'Map the Value-Stream Principle' requires good understanding of three disciplines: the system domain, Systems Engineering, and Lean. It also needs devoted leadership, planning expertise, patience, and sufficient time and budget to create a plan properly. Good planning can pay out tremendous dividends in subsequent program cost savings, schedule shortening, and value quality. Therefore, mapping must not be rushed. The learning curve is steep, but, once learned, each next program mapping becomes vastly easier. Section 4.2 provides some guidance.

The ease or difficulty of implementation of the 'Flow Principle' depends critically on the clarity and stability of a value proposition and the degree of detail of Value-Stream Mapping. If both are well done, then robust, predictable, and efficient flow is much easier to achieve. In a well-functioning Lean program, everybody notices when flow makes good progress and when it stops, without using exotic and expensive metrics. Making a complex program flow requires a number of supporting activities and characteristics to be in place, such as good leadership and management; availability of expert knowledge of the domain and of systems engineering; teamwork and alignment of stakeholders for the program goals and value; excellent communications and coordination; real-time visibility of status, issues, and issue resolution; and excellent partnering relations with suppliers and other stakeholders.

In manufacturing, the 'Pull Principle' and the closely related Just-in-Time (JIT) production and *Kanban* signals are some of the most difficult to implement. In contrast, in PD environment, JIT is not as demanding: Tasks must be completed before their outputs are needed (too early completion may make the output obsolete and too late causes delays), but the exact timing is not as strict. Instead, solid planning of task precedence is needed, as well as excellent coordination of tasks with data users for *right the first time* execution. These are relatively easy to put in place.

The 'Perfection Principle' necessitates a culture of making imperfections visible and the motivation and infrastructure for continuous improvement of processes. It also requires a comprehensive infrastructure for a knowledge database, lessons learned, and education.

The 'Respect for People Principle' may be the most difficult to implement if the enterprise starts from a traditional authoritarian, *stove-piped,* and bureaucratic culture. The transformation requires solid knowledge of Lean, excellent leadership, corporate support, training of the workforce, mentoring, and lots of good examples.

As mentioned in Chapter 2, Lean Systems Engineering adopts a number of practices previously known under other names, such as Six Sigma; Quality

Management; Design for Manufacturing, Assembly, and Testing; *test as you fly*, and others. The criterion we use for adoption is very simple, stated in the following box:

> If a practice promotes value, reduces waste, and can be described by the six Lean Principles, it is called Lean and adopted here.

Chapter 4

Lean in Product Development

There is a vast productivity reserve in Product Development programs, averaging 88% of charged time. This is our opportunity to make significant progress using Lean Thinking. - as explained in this Chapter

4.1 REVIEW OF PROGRESS

Lean Product Development (LPD) is a fast-growing field, driven by the competitive global need to develop products and systems better, faster, and cheaper. It is also driven by the dynamic creativity of engineering, the profession of unrelenting problem solvers.

Toyota is widely recognized as the leader in LPD. Its practices have been described by a number of oft-cited articles and books [Kennedy, 2003; Liker, 2004; Liker et al., 1996; Sobek, 1997a, 1997b; Sobek et al., 1998, 1999; Ward et al., 1995a, 1995b], and the comprehensive book by Morgan and Liker [2006], and are not repeated here, although some enablers listed in Chapter 7 have been inspired by these.

Clark and Fujimoto [1991] published a comprehensive approach to PD, introducing strategic metrics of automotive enterprise performance—lead time, quality, and productivity. They described PD project integration, strategy, planning,

Lean for Systems Engineering with Lean Enablers for Systems Engineering,
First Edition. Bohdan W. Oppenheim.
© 2011 John Wiley & Sons, Inc. Published 2011 by John Wiley & Sons, Inc.

management complexity, integration of problem-solving cycles, and the Toyota and Honda project management styles, which they describe as "heavyweight."

The need for Lean Thinking in defense PD programs was manifested by several Lean Advancement Initiative (LAI) studies of waste.[1] They are discussed at length in Murman et al., [2002] and Oppenheim [2004] and are only briefly summarized here: The amount of direct waste (the waste of charged time to produce the contracted value) in government PD programs has been estimated between 60 to 90% of charged time, with about 60% of all tasks idle at any given time [Browning, 1998 & 2000; Chase, 2001; Joglekar and Whitney, 2000; McManus, 2004; Millard, 2001; Young, 2000]. According to these authors, while such estimates arguably lack scholarly rigor, they are consistent enough across corporations, programs, and years to yield a comfortable level of confidence. This extent of waste implies vast productivity reserve in PD, averaging 88% of program charged time, and thus an opportunity to make significant progress using Lean Thinking. Oehman and Rebentisch [2010] created a white paper entitled *Waste in Lean Product Development*, published under the new LAI Series: "Lean Product Development for Practitioners."

When discussing waste in PD programs, we should point out that the present book is limited to the analysis and removal of waste in already contracted PD programs. This book does not deal with the societal waste of unneeded programs forced onto the defense community by aggressive marketing or congressional pressures, programs that have been closed for non-performance or for political reasons, less-than-perfect government acquisition regulations and related politics, the tortuous process of getting a program funded, or government bureaucracy. Also ignored is waste in early PD program phases before Milestone A, such as the convoluted process of identifying potential enemy threats and capturing related requirements; the bureaucracy of program approvals by numerous and disjointed power centers; etc. All these waste topics deserve a comprehensive Lean treatment and possibly a separate set of "Lean enablers for acquisition.[2]" But the reader will not find them in this book, except marginally as they relate to program execution.

In the 2000s, important contributions to LPD were made by individuals associated with the LAI community. McManus [2004] published a manual for PD Value Stream Mapping. Oppenheim [2004] reformulated the Toyota manufacturing approach to the field of product development using the Lean Principles (summarized in Section 4.2). This approach is suitable for low-risk smaller programs or for individual milestones of a larger program. Rebentisch [2005] presented a seminal

---

[1]The PD program milestones established by DoD Instruction 5000.02 [2008] are: Milestone A, which approves entry into the Technology Development phase; Milestone B, which approves entry into the Engineering and Manufacturing Development phase; and Milestone C, which approves entry into the Production and Deployment phase. The sources quoted above measured the 88% waste in post-milestone B programs.

[2]When this book was in final stage of production, Lean Advancement Initiative at MIT initiated a new effort of developing "Lean Enablers for Program Management", using the present "Lean Enablers for SE" as the starting point. The new effort is a cooperative project between INCOSE, PM Institute, and others. The public website is: http://www.lean-program-management.org/

lecture on LPD that became a starting point for several subsequent studies of the field. Rebentisch and McManus [2007] developed a teaching simulation for Lean Enterprise Product Development, as well as a tutorial on LPD of complex products. Murman [2008] and Haggerty and Murman [2006] comprehensively describe the use of Lean in specific aerospace engineering programs, although most of these programs were not set up for Lean and demonstrated strong Lean characteristics only in post-analysis. Lempia [2008] and Egbert et al. [2008] describe two successful PD programs in aerospace environments that were set up using Lean principles. Reinertsen [2009] published a comprehensive book formulating 175 principles on Lean PD, most of which are applicable to both commercial and government programs.[3]

Private contacts of the author with industry in the United States, Europe, Israel, and Asia indicate that many more organizations are attempting to implement Lean in PD, but have not published their results to the author's knowledge.

So, overall, how mature is the field of LPD? An important question that PD managers ask themselves is: Is LPD knowledge mature and practical enough to be applied to real PD programs? This author is convinced that the answer is emphatically positive. As a profession (or more precisely, the fraction of the profession familiar with Lean), we believe that we understand which management and engineering practices increase customer value, reduce waste, promote shorter program schedules, lower costs, and lower stakeholder frustrations. Early case studies, presented in Section 8.2 of Chapter 8, even though based on fragmentary implementations of Lean, and some lacking scholarly rigor, indicate excellent benefits. We understand the corporate-level preparations of people, processes, tools, and infrastructure that are necessary to support efficient programs. We understand excellent techniques for program planning. We understand the practices that promote predictable flows of value and those that slow progress. We understand wastes, and we are quite good at identifying them. Finally, we also understand the cultural aspects that are conducive to Lean, and those that resist Lean. We understand some domains better than others. For example, Lean automotive PD is understood better than large governmental weapon-system programs, which are more immune to competitive pressures for efficiencies. Overall, qualitatively, we understand enough about Lean PD to use it routinely. Because precise LPD ROI data is not yet available, should we wait and not attempt LPD? The author suggests that this question is similar to the ones about smoking or diet: We should heed the advice, "do not smoke," and "follow a healthy diet," even if we are unable to predict accurately the consequences of good habits for an individual. So, LPD practices are worth pursuing even if we cannot accurately predict the savings or the ROI for a particular program being planned. Noticing PD waste and not doing anything about it should not be an option. At this point in the development of technology and civilization, LPD represents the state of the art in system development. With unlimited human

---

[3]A frequent reaction of Systems Engineers to 147 Lean Enablers for SE is that "this is too many for Lean." The fact that Reinertsen lists 175 Lean principles provides support to the argument that the complexities of PD and SE require a comprehensive list of practices, rather than a few slogan-like statements. Slogans may yield some awareness, but cannot be expected to significantly improve programs.

Why are they immune? No incentive for competition — No incentive for profit over-regulation

creativity and entrepreneurship, new PD paradigms will probably appear, but at this time Lean is state of the art.

As just noted, LPD is far easier to implement in commercial programs than in governmental programs. The latter involve a huge and rigid acquisition system bureaucracy and are driven by incentives that often oppose Lean, such as *cost-plus* contracting, where contractors are paid for effort plus a fee; political and contractor lobbying; geographical distribution, which stifles program co-location and coordination; lack of real competition between large contractors; exceedingly long program durations that lead to costly requirement changes; unstable funding; difficulty to formulate good requirements in a rush to get programs underway; excessive and unstable requirements and runaway requirements changes; short-term Wall Street pressures; dissolved management and distributed subcontractors without centralized Systems Engineering; and many others. However, as pointed out above, even some recent military PD programs have demonstrated Lean characteristics with excellent benefits. In summary, this author recommends an aggressive implementation of LPD in all programs, in all domains, and in all environments.

One thing is certain: In the long term, operating in global competition, we have no choice. If we procrastinate and fail to implement competitive LPD, the dynamic global competition will not wait for us. Even military programs, with their increasingly global supply chain, are no longer totally immune to global competition. The morgue of industry is filling up with the companies that became stubbornly complacent and procrastinated too long. The choice is ours to make.

4.2 THE METHOD OF LEAN PRODUCT DEVELOPMENT FLOW (LPDF)

This section describes a holistic process called *Lean Product Development Flow* (LPDF) intended for smaller programs or program segments. LPDF represents an adaptation of Lean manufacturing methodology to PD environment. It was created as a contribution to Lean Systems Engineering, but it is an earlier and separate work from the *Lean Enablers for Systems Engineering*. However, several of the Lean enablers listed in Chapter 7 have their origin in LPDF. The following text has been adapted from a full-length article [Oppenheim, 2004].

4.2.1 Introduction to LPDF

The LPDF method is intended for a super-efficient and fast execution of programs and projects while promoting best value. It does not, however, apply to all programs. The following box defines the limitations:

> LPDF is limited either to those smaller programs that involve only mature technologies and low risks and are understood well enough so that a detailed plan can be created for the program at its beginning; or to one or more program segments that can be characterized as such.

In this chapter, we use the term *project* to denote such a small program or a program segment. LPDF project may involve limited-scope risk or research, provided it can be identified early in the project and carried out separately from the critical path.

LPDF execution is organized as a rhythmic work flow through a series of short and equal duration *takt periods*, each period ending in a comprehensive meeting called an *integrative event*. The Event is devoted to structured, comprehensive coordination and resolution of all issues that come up during the *takt period*.

LPDF requires solid preparations of people, detailed and competent planning of the LPDF process, and disciplined execution. These preparations include selecting a Core Team of managers (or highly competent deputies), individuals representing all of the applicable functions and major stakeholders, and training of teams. Planning involves detailed Value-Stream Mapping, parsing of the Value Stream map into equal *takt periods*, and architecting the LPDF team, if necessary, using dynamic allocation of resources. LPDF also requires an excellent leadership style, which is modeled after the Chief Engineers of Toyota, Honda, and early U.S. aerospace programs. A project Chief Engineer must be expert in the project domain and have a good understanding of at least the first-level interfaces and the tradeoffs between all major subsystems. The individual must be an expert systems engineer and a strong leader who is skilled in consensus building, while also focused on the integrity of both the project and system (or project deliverables, if this is only a segment of a larger program). The Chief Engineer is responsible for the entire project and is accountable for and has authority over both the technical and business success of the system being designed. Assistant Chief or Chiefs assist in selected technical areas. A Project Manager reports to the Chief Engineer and assists with project administration and finances.

4.2.2 Lean Manufacturing: A Refresher

LPDF shamelessly steals some concepts from Lean Manufacturing. Therefore, it is useful to review that field. In manufacturing, the term *takt time* denotes the amount of time allocated to each workstation on the moving line, which allows for a robust completion of its task so that the entire line can advance one station at a time. It is also equal to the rate at which finished products come off the production line and are shipped to customers and synchronized with customer orders. Each worker or process must work at a common *takt time*; otherwise, pileups or gaps occur before and after the offending workstation, and the line must then stop to catch up. As customer orders increase, production rates and line speed must increase, while *takt time* decreases. This is accomplished by adding resources—up to the capacity of a system—rather than by forcing processes to run faster. The capacity must be realistic: No worker should be asked to work faster than the well-tested ergonomic rate. This realistic capacity is made possible by worker training and machine and process optimization. If this process is not followed, workers will not be able to complete their tasks and inevitably will produce defects. Flexibility in adding and removing human and machine resources is an important factor in Lean system profitability. The term *flow* denotes the uninterrupted or pulsed motion of work pieces at a steady pulse of **takt time** through all processes with no backflow or rework.

After more than a century of using production lines in factories, we tend to take them for granted. Yet, making complex production line flow according to the **takt** pulses is quite difficult. Implementation requires carefully splitting and balancing total assembly work among individual lined-up workstations or work cells, perfecting each process, and providing each worker with adequate parts, tools, training, and ergonomics to make timely and robust completion of any task possible. Numerous references describe this production system [see, for example, Spear and Bowen, 1999].

> The key to success, which is also the key to the LPDF method, is the ability to plan and parse the total work into tasks of equal duration and size, so that each task becomes predictable in terms of outcome, quality, effort, and *cycle time*.

Lean production is the most efficient method known for flexible delivery of quality products in the shortest possible time and at minimum cost, with minimum inventory and perfect quality. Henry Ford was the first to line up machines in sequence and split the standard work among workstations into equal-duration tasks. It was Toyota, however, that invented the most advanced Just-in-Time system with minimum inventories; a work culture based on teaming, openness, empowerment, and trust unmatched by its competition; and an amazing quality system that obviates the need for final inspection [Liker, 2004].

4.2.3   Overview of LPDF

Figure 4.1 schematically illustrates the LPDF on the project timeline.

The project effort begins with a precise value definition: We must fully understand what the project is to deliver before we can embark on a detailed planning of it. The LPDF planning may be impossible if value is poorly formulated, unstable, lacking clarity, completeness, comprehensive capture of need, or consensus.

Once the value proposition is fully defined, detailed planning using the Value Stream Map shown in Figure 4.1 can begin.

The flow can begin after the Value Stream Map (VSM) is completed. The flow ends with the release of the value deliverables. Again, if the project is a complete program, it delivers a created system design ready for production and approved by the manufacturing stakeholders. If the project deals with only a segment of a larger program, the deliverables are information packets that feed the next program segment.

Between the two ends, flow proceeds at a steady pulsed rate, as on a moving line, as follows. The flow consists of a sequence of equal **homework** periods called

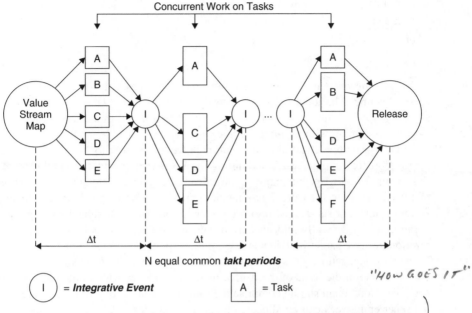

Figure 4.1 Schematics of Lean Product Development Flow [Oppenheim, 2004]

takt periods (of equal duration Δt), each terminating in an *integrative event* (I). In manufacturing, the term *takt time* denotes the rate at which finished products come off a line. In a project, we are typically creating a unique system, so it makes no sense to speak of multiple products. To emphasize this distinction, we use the term *takt period* rather than *takt time* to designate the equal-duration time periods used in the project. *Takt periods* of very short duration are recommended, typically one week long, to expose any and all issues in real time and enable immediate and effective issue mitigation while a problem is small and easy to deal with. The *takt periods* provide a constant and frequent rhythm and common disciplined deadlines for all tasks to the entire project team.

Within each *takt period*, engineers work on parallel (concurrent) tasks, denoted A, B, C, and so on. A task is defined as a work element that has clearly defined major inputs and outputs and that can be executed as a self-contained activity independent of other tasks. Typically, such tasks represent much smaller work packages than those used in the Systems Engineering Work Breakdown Structure. The individual letters denote tasks executed by different major functions (departments), or individuals, or major suppliers involved in a system creation. For example,

A = architecture
B = 3D design
C = modeling
D = dynamics

E = a major subsystem

F = test

etc.

All **takt periods** are of equal duration, but not necessarily of equal effort or person-loading. The ideal LPDF is a steady progress of the value stream through all **takt periods**, with frequent maximum coordination and minimum waste, each **period** terminating with an **integrative event**.

LPDF places few constraints on team architecture, except for an absolute need for the Core Team to be co-located during the VS mapping, and during all **integrative events**. Employees and major suppliers can be organized into any configuration that makes sense to the Core Team, including into long-term and short-term teams, system teams and sub-teams, into groups or as individuals assigned for a project duration, or dynamically allocated from their home departments in a matrix organization, or as separate individuals (e.g., hired experts or supplier representatives). In matrix organizations, it is important to have department heads evaluated, among other factors, on the basis of the degree of support they provide to the Chief Engineer. The Core Team should comprise the same individuals for a project's duration, but other employees can be allocated flexibly, as needed.

All individual tasks have the same duration but involve different amounts of effort. Depending on the amount of work needed to complete a given task in a given **takt period**, the staffing for tasks will vary accordingly. Some tasks may need only a single individual, others a small team with a mix of full- or part-time employees, or a subcontractor. Some tasks may not be needed at all in some **takt periods** (e.g., Task B is missing in the second Period in Figure 4.1). In order to comfortably function in this work environment, all employees should be trained for the special needs of LPDF.

Even though tasks are self-contained by definition, in PD projects various questions come up all the time and they should be resolved immediately without waiting for the next **integrative event**. Each employee who has a question should immediately locate the person who can provide the answer and seek that answer without wasting time. Thus, besides structured communications during **integrative events**, informal communications take place all the time. Therefore, all stakeholders should have a good access to a directory and organizational chart of the entire project team. The infrastructure should be organized to facilitate such informal communications.

As many tasks should be concurrently carried out in any given **takt period** as possible. Concurrent execution is the most effective tool for shortening overall project time, but it requires detailed planning of the task scheduling and precedence, and their inputs, outputs, and control points. The "Critical Chain" formulated by Goldratt, [1997] explains how to estimate task times without padding. In order to compensate for the accumulating task completion variability, it may be prudent to insert time buffers into the VSM [Goldratt, 1997]. The buffers assure that the overall schedule will be met even if some individual task exceeds its planned duration.

Given the intrinsically tight schedule of LPDF, robust and timely completion of tasks is critically important. Chief Engineers should insist that failure to meet ***takt period*** deadlines will be justified only as a rare exception for only incontrovertible reasons. Tasks should be staffed and people trained and incentivized accordingly. The role of the LPDF Core Team is to provide the resources, coordination, and training required to make that possible. The Chief Engineer and his or her staff, as well as Core Team members, should be available for guidance, mentoring, even ad-hoc training, if needed, as well as general support for flow.

4.2.4 Integrative Events

An ***integrative event*** is a meeting where all current issues are openly discussed, mitigated, and the work comprehensively coordinated, verified for consistency with value proposition, and prepared for the next ***takt period(s)***.

In contrast to the spontaneous communications taking place during ***takt periods***, ***integrative events*** are strictly scripted. The Chief Engineer, or an assistant, should define the ***integrative event*** agenda and structure and lead the meeting. Box 4.1 provides a checklist for the ***integrative events*** [Oppenheim, 2004].

Box 4.1 *Integrative Event* **Checklist [Oppenheim, 2004]**

a. Efficient review of progress. Chief or Assistant Chief asking pointed, knowledgeable questions of participants, including numerous questions "why?" asked in a non-confrontational style.

b. Comprehensive coordination of work.

c. Resolution of tradeoffs, concerns, issues, and building consensus—if practical, in breakaway sessions, involving only the needed individuals.

d. Identification, management, and retirement of project risks.

e. Exploration of design spaces versus point designs.

f. Optimization and coordination of the inevitable iterations for minimum effort and cost.

g. Decisions whether to insert knowledge from legacy projects.

h. Involvement of suppliers and other stakeholders.

i. Reuse of modular subsystems and checklists from former projects.

j. Balancing tradeoffs between design margins and the analysis fidelity.

k. Discussion and decisions on which analysis, tests, and artifacts are needed, resisting those deemed wasteful.

l. Adjustments of VSM, assignment of adjusted work to responsible parties, and allocation of necessary resources.

m. Addressing any and all big relevant questions.

n. Gate-based reviews, if applicable.

It is recommended that *integrative events* be held on Fridays. At the latest, on Thursday afternoons, by a set time (e.g., 15:00 hours), the office of Chief Engineer (and key stakeholders of an issue) should be notified of any issue that is to be scheduled for a discussion on Friday. All cross-functional issues, potential delays, potential baseline changes, risks, cost variations, non-conformance, or other issues should be reported to the Chief's office. The reporting should be done as efficiently as possible. Standardized so-called A3 forms, as the only attachments to emails, are recommended for such notifications.. Subenabler 3.5.6, defined in Chapter 7, describes this form. The Internet contains a wealth of examples of A3 forms. It is recommended that all project stakeholders be trained in the use of the preformatted A3 forms and discouraged from writing verbose memos, emails, and reports. Engineers tend to lack good writing skills and consume a significant fraction of precious program time in writing, which tends to be imperfect. A pre-formatted A3 form with only salient bulleted statements saves time in writing and reading and transmits information concisely. An additional benefit of the A3 forms is that they provide an excellent and easy-to-catalog record of the project and an easy means for sorting the issues. If the recipient of an A3 form needs more information, he or she should ask for it explicitly. On Thursday afternoon, after the deadline, a staff person in the office of the Chief Engineer should sort received A3 forms into an agenda to be followed during the next day's *integrative event*. The Event should continue (if necessary, spilling into the weekend) until all issues are resolved and all questions addressed, so that the next week's schedule is not affected.

It is obvious that the scope of *integrative events* extends well beyond the frequent practice of status reviews. The Chief Engineer should frequently question engineers as to the "why" of their decisions and practices to stimulate better decision making and to verify correctness.

Co-located meetings of the Core Team (supported by staff as needed) in a common *war room* (a.k.a., project room) are absolutely critical. Strict policies should be in place regarding the overdissemination of documents intended for the *integrative events*, otherwise team members may become overwhelmed with pushed data. Strict discipline for perfect attendance is recommended.

The *integrative events* may be recorded to enable easy recall and to augment corporate memory for future PD projects. Video recording may be used as an efficient record-keeping device for memory jogging, provided strict rules are in place prohibiting any use of the video for staff evaluations, contractual compliance, or other such abuses.

4.2.5 Selecting the Project Schedule

The total LPDF time (a.k.a. project *throughput* time, or *cycle time* or schedule) is a critically important part of the value proposition. Yet, in practice, it is an arbitrary aspect of PD. The time must be decided at the beginning of the Value-Stream Mapping effort for the same reason as for a moving manufacturing line. Ideally, the *throughput* should reflect the time when the stakeholders need the deliverables, or the time needed to beat the competition, rather than the schedule convenient for the team. In the absence of a stakeholder-set deadline, a radical step is recommended here to reduce the traditional legacy-based (or proposal-quoted) *throughput* time by the fraction of the PD waste that the project management is ready to tackle. For example, if management estimates that the amount of self-evident and easy-to-eliminate waste on a recent similar project was 20%, the *throughput* time for the current LPDF project should be cut by that fraction. Ambitious leadership might favor more aggressive cuts. (After five years of experimenting, Henry Ford realized a 90% *throughput* reduction upon evolving to his assembly line from the former craft production, although it is doubtful that he could have predicted this *à priori*). The risk in schedule cutting is small: At the worst, the schedule may slip back toward the traditional asymptote. In the spirit of continuous improvement, larger cuts should be possible as experience with LPDF increases. Admittedly, this is a radical and arbitrary approach. However, it is not less arbitrary than the current practice of proposal teams cutting 25–50% from the schedule estimate to compensate for suspected padding.

4.2.6 Mapping the Value Stream

Common wisdom calls for good planning at the beginning of PD projects. Experience-based, competition-motivated, consensus-created streamlined Value-Stream Mapping (VSM) parsed into short *takt periods* is the ultimate good plan. The VS must be mapped before the LPDF flow can begin. While subsequent execution permits flexible adjustments of tasks in real time, the adjustments should be used as a tactical mitigation of issues, rather than a poor substitute for good initial planning. The VSM must list all tasks that contribute to value, starting with project input requirements and other descriptions of need and ending with value deliverables. The map combines a process map with data about each task, indicating the effort and *cycle time* data (the task *cycle time* does not need to be equal to the common *takt period* yet; that will come later). The VSM is a comprehensive planning effort.

Value-Stream Mapping of LPDF consists of the three following steps:

a. Current-State Mapping
b. Future-State Mapping
c. Parsing the Future-State Mapping into *takt periods*

These steps are discussed in turn.

Current-State Mapping   The Current-State map is a detailed graphical representation of the PD plan using current project methodology (before any streamlining), showing all tasks, their precedence, and control points. This is a starting point for subsequent identification of waste and streamlining. If available, the Value-Stream Map actually used on the most recent similar (or legacy) project, with its task precedence and lessons learned, is an excellent way to begin the Current-State map. Good corporate memory is invaluable in this step. Incentives should be introduced to capture and preserve the Value-Stream Maps from all projects (as well as the various checklists, system and component performance charts, trade-off charts, non-dimensional ratios, architectural properties, and numerous other useful design-capture data). Starting with legacy knowledge is helpful even if only partial information is available, such as the process map of an actually executed project or the Gantt chart. Participation of a high-level manager from that prior project in the VS mapping of the current project is highly desired to indicate the dos and don'ts to the new team.

Each task should be described on a separate task sheet, standardized with the following fields:

1. Task number and the week of execution (left blank until Future-State Mapping).
2. The person responsible (name, title, telephone, cell, email, location; again left blank).
3. Major inputs, each indicating the source tasks.
4. Major outputs, each indicating the destination task and approval or control nodes.
5. Brief description of effort and scope.
6. Warning about potential issues; any other notes or comments.

If possible, each task sheet should be temporarily placed in the best-guessed week in which it was executed on the legacy project, or where the task "owner" feels it should go. The idea is to include all tasks, and then iterate their placement in the timeline later. Where available, sticky notes should indicate identified waste for subsequent removal, e.g., the time of waiting for or chasing data, time wasted waiting for signatures, rework, *reinventing the wheel*, etc. Next, proper precedence of tasks should be determined, usually by iterations. This iterative process may appear messy, but it offers huge payback potential in the Future Step mapping.

Next, or concurrently, tasks from a legacy project should be tailored, amended, and modified to reflect the current project. From this point on, the effort should be handled by a complete Project Core Team comprising experienced functional managers (or competent deputies) representing all major subsystems, representatives of major suppliers, and the Core Team leaders (Chief Engineer, Assistant Chief Engineers, and Project Manager). If the team writing the proposal is different from the Core Team, the former should be represented during the mapping effort. A typical Core Team may involve 10–15 individuals. Each Core Team member is free to bring along any staff person for help, but the staff should not release the Core Team members from making all major decisions, negotiating, iterating, and reaching consensus.

Brainstorming, negotiating, and iterating are the most productive means at this stage. The Chief Engineer experienced in PD VSM and possessing good motivational and leadership skills should lead the effort. The present focus should be on listing 'all' tasks and their waste, rather than on any task or flow optimization, which comes later during the Future-State Mapping.

Future-State Mapping   This step has a potential for huge direct ROI, often measured in millions of dollars saved per hours of effort. Therefore, this mapping step should be performed as comprehensively as possible.

The Core Team may conclude that the project involves one or more uncertainties that would pose a risk to the project schedule if left within the main workflow. Each such uncertainty should be isolated from the main flow, placed on a separate track, assigned to a separate sub-team, and staffed to resolve the uncertainty in time for deployment in the main flow.

The Current-State map, displayed on the walls, becomes the basis for the iterative waste removal and for improving task concurrency, synchronicity, precedence, and general flow. Experience indicates that some NVA is self-evident and easy to remove, some NVA and RNVA will require brainstorming and negotiations within the Core Team, and some may be discovered only in a future project, in the never ending pursuit of waste removal and streamlining.

Parsing the Future State Map into Takt Periods   The last phase of VSM is to parse the Future-State Map into *takt periods*. This step and the previous step of Future-State Mapping may require iterations together. Again, the entire Core Team should participate to enable iterative brainstorming and negotiating. The dynamic allocation of employees during different *takt periods*, if practiced, should be addressed at this stage. The parsing may open additional opportunities to remove waste and to optimize task precedence and flow.

The application of fixed-length *takt periods* is an absolute requirement for disciplined flow, just as a manufacturing moving line must involve work packets of equal *takt time*. All work must be parsed into equal *takt periods* (lasting just short of four days of work to reserve a few hours on Thursday afternoon to report issues to the office of Chief Engineer and Friday for the *integrative event*). The parsing must be done either by formal splitting of longer tasks into smaller ones, or by logical splitting of such tasks for reporting purposes. For example, a task inherently longer than the *takt period* (e.g., a multi-week test) can always be logically subdivided into shorter *takt periods*, so that the engineer responsible for the task can report during the given *integrative event* that "the task number ... proceeded as planned during the *takt period* number...," or: "the following issues have already been identified in the task...."

Clearly, in any complex flow involving many stakeholders, unexpected events, uncertainties, and design changes may occur, requiring adjustments to the schedule, as they do on automotive assembly lines. The general attitude of the Core Team should be to map the best Value Stream possible, but also to prepare for flexible handling and mitigation of the changes. PD experience indicates that an imperfect plan is better than none.

The role of the Chief Engineer is to guide the Core Team toward consensus on the VSM. The mapping should continue until that goal is met, that is, until every Core Team member accepts the final parsed VSM and declares readiness to provide the required resources and complete tasks when planned.

Detailed VS mapped into short *takt periods* at the beginning of the project automatically constitutes a detailed project plan and schedule. Theoretically, the subsequent monitoring of project progress could be as simple as checking off task sheets in the VSM displayed on the walls. This reduces the need for complex metrics and costly bureaucracy monitoring of the progress.

4.2.7   Project Leadership and Management

Good leadership cannot be delegated or automated. A highly skilled leader, named the Chief Engineer, modeled after Toyota [Sobek et al., 1999], Honda [Clark and Fujimoto, 1990], and the Skunk Works model [Rich and Janos, 1994], should lead the entire LPDF project. The person's job description should be to *produce the required product (or deliverables) to the satisfaction of the stakeholders, within budget and schedule*, and the person should be evaluated only by how well this goal is met. The Chief must be the sole *project owner*, totally responsible, with authority and accountability for the entire project (preparations, planning, concepts, tradeoffs, key design decisions, coordination, targets, schedule, and budget.) The Chief should be ultimately responsible for balancing the technical success with the business case. (See also enabler 5.5 in Chapter 7.)

Box 4.2 contains a summary of the desirable attributes of the Chief Engineer, Assistant Chiefs, and administrative Project Manager.

The company involved in LPDF projects should groom several Chief Engineers for each major product domain, support their professional growth and education, expose them to challenging experiences, and rotate them through major departments. Candidates should be carefully selected from among the brightest and the most promising, both technically and for their interpersonal skills. Candidates should prove themselves as Assistant Chiefs before being promoted to the Chief's position.

Historically, aerospace and defense programs used the equivalent of a Chief Engineer. Examples abound: Jack Northrop in the early Northrop company; Howard Hughes (before his powerful mind was overtaken by illness); Kelly Johnson or Ben Rich at Skunk Works [Rich et al., 1994]; leaders of the early NASA's Mercury, Gemini, and Apollo programs [Johnson, 2002]; Admiral Hyman Rickover of the U.S. Nuclear Submarine program [Rockwell, 1995]; and numerous others. In these programs, nobody was ever in any doubt about who was in charge of the program, and it was always a highly competent engineer and leader, never a financial manager enslaved to Wall Street expectations. These people led the development of highly successful systems, and achieved business success at the same time. The unfortunate recent practice in defense programs has mostly abandoned the Chief Engineer position, dissolving the responsibility among the Program Office

Box 4.2 Desirable Attributes of Chief Engineer, Assistant Chief Engineers, and Project Manager [Oppenheim, 2004]

Chief Engineer

- *Interpersonal skills*. Ideally, a good leader, with high degree of credibility, who is free of a domineering personality. Leading and motivating for excellent performance using non-confrontational style. More like a movie director or symphony conductor than a drill sergeant. In frequent personal contact with engineers, but without micromanaging. High level of interpersonal skills to guide the team toward consensus during the value proposition and VSM work, when resolving issues during *integrative events*, and when negotiating with the company for resources. Ability to draw on team members' competence, experience, and creativity.

- *Education*. Preferably a master's degree in Systems Engineering, with significant experience in the domain, or a master' degree or equivalent in the product domain, with at least several courses in Systems Engineering.

- *Experience*. Solid understanding of all critical first-level subsystems, their interfaces, tradeoffs, and risks. Experience in the capacity of an Assistant Chief Engineer on at least a few programs. Knowledge of frustrations, problems, and solutions experienced in former programs. Understanding of company culture. Preferably most professional years spent rising through the ranks and rotating through major departments as an active engineer. Record of lifelong learning, attending professional conferences, and following literature.

- *Freedom of Action*. The Chief must have the freedom to select a few Assistant Chief Engineers to complement the Chief's expertise, whose loyalty is to the Chief, the end customer (i.e., the project), and the company, and not to any particular functional department from which they came. The Chief alone should evaluate the Assistant Chiefs. Also the freedom to execute LPDF according to Chief's own preferences.

- *Focus*. Never-ending focus on customer satisfaction, other stakeholders' needs, project value and integrity, and reduction of waste.

- *Compensation*. Clearly, the Chief's compensation should be proportional to the exceptional role the person plays and the vast responsibility.

- "The most coveted job in the company."

Assistant Chief Engineer

An experienced and competent engineer showing promising skills in all above areas but lacking the extensive experience required for the Chief's position. Chief Engineer selects one or more Assistant Chiefs from the small pool of candidates.

Project Manager

Project Manager reports to the Chief Engineer. The Manager understands and tracks in real time all project costs and provides real-time accounting support to the Chief Engineer for his or her decisions. The Manager also handles administrative support for the program in order to free the Chief Engineer to focus on the value creation. However, it is the Chief who balances the technical and the business success of the project.

team, which has had a typically weak and financially focused Program Manager; a narrowly focused Systems Engineering lead who lacked overall responsibility for the program success; weak IPT managers; and territorial managers of engineering departments; and subcontractors, with nobody truly in charge of the entire program. These practices should end as soon as possible.

4.2.8 Project Room

The VSM is a complex graphic even in moderate-size projects, so a large room with ample wall space (colloquially called the *war room*) should be dedicated to a project for its entire duration. The VSM planning effort, all *integrative events*, and ad-hoc meetings should be conducted in this room, with the VSM and project notes conveniently in view. Computer screens are strongly discouraged; they are too small and too difficult to see the entire project map. Wall layout is preferred to an electronic implementation, because it enables Core Team members in real time to read all tasks and brainstorm and negotiate task parsing, precedence, concurrency, synchronicity, scope and effort, inputs, outputs, and waste in order to reach a consensus.

Takt periods should be delineated on walls by vertical marks for an entire project duration, and for easy posting of each task sheet in its designated week. Each task sheet should fit into an A4 (letter-size) page, thus the vertical lines should be spaced apart by about 30 cm (about one foot), requiring a room of about 15 m (about 50 ft) of clear wall circumference for a one-year project. Oppenheim [2004] articulates details on the recommended room architecture. Ideally, a few smaller rooms should be available nearby for breakaway discussions. The offices of the Chief Engineer, Assistant Chiefs, and Project Manager and their staffs should be located in close proximity. The room should contain networked computers, printers, projectors, ample writing materials, and a large conference table with enough chairs to accommodate the Core Team.

4.2.9 Closing Remarks about LPDF

Disciplined work execution within short *takt periods* is an important element of LPDF. Compelling arguments favor this approach. If not followed as recommended here, the penalty to the project would be less-than-full benefit, but hardly an increased risk of loss of system integrity. The resultant penalty in cost and schedule should not be worse than that of recent traditional programs. In other words, LPDF offers potential for radical benefits, with no cost or schedule risk beyond those of traditional programs.

LPDF represents an adaptation of Lean manufacturing to project environment. Over the last century, significant knowledge, experience, and effort have been devoted to the design of flow in automotive lines, and we tend to take that progress for granted. Yet, even today's best assembly lines still suffer from frequent stoppages due to unexpected problems. The author observed a Toyota line in the NUMMI plant in Fremont, California, recognized as one of the best in the world, stopping several times per hour while assembling a mature model of Corolla™. Local employees confirmed that this frequency of stoppages was normal. This demonstrates that problems are to be expected even after a significant experience. Therefore, it would be naïve to expect no problems in a LPDF flow. Nevertheless, the significant potential benefits and minimal risks make the method worthwhile.

Chapter 5

From Traditional to Lean Systems Engineering

"Traditional Systems Engineering is a practice which has a lot of strengths, but is not as good as it could be."

—Lean Systems Engineering Working Group of INCOSE

5.1 SUCCESSES AND FAILURES OF TRADITIONAL SYSTEMS ENGINEERING

Human civilization has successfully used systems thinking to create many complex systems such as the pyramids, Roman aqueducts, and, more recently, infrastructure, communications, phone networks, health, education, safety, emergency response, war effort, and many others—all without using the formal term Systems Engineering (SE). SE applies to all engineering domains. The practice of SE is now mandatory in U.S. government technology acquisition programs. The use of SE in commercial programs is voluntary, but increasing.

The mandated use of SE in government programs is a relatively recent U.S. government policy—only announced in 2000. For several years prior, NASA operated under the so-called *Faster, Better, Cheaper (FBC)* policy, which the U.S. Department of Defense termed *Acquisition Reform (AR)*. Both policies were blamed for

Lean for Systems Engineering with Lean Enablers for Systems Engineering,
First Edition. Bohdan W. Oppenheim.
© 2011 John Wiley & Sons, Inc. Published 2011 by John Wiley & Sons, Inc.

numerous system failures during the 1990s, adding up to $12 billion losses in space systems.[1]

The subsequent study sponsored by the U.S. Congress [Young, 2000] diagnosed that FBC/AR removed much of government oversight, made prior mandatory standards optional, and permitted contractors to cut Systems Engineering efforts as well as tests. These cuts, according to the study, led to the failures. The report recommended a stronger role of SE in programs, a return to more oversight, standards use, and testing *as you fly*. It also recommended more relaxed budgeting and *cost-plus* contracting. These recommendations were adopted, and the number of program failures decreased dramatically. Regretfully, the technical successes in the post-FBC/AR period were not accompanied by business successes. Abundant evidence indicates Systems Engineering was performed less then satisfactorily in the business sense, with many recent programs experiencing severe budget and schedule overruns, with final budget and schedule growing significantly beyond those of earlier periods, and some programs being closed for non-performance, as described in Section 5.2.

Overall, the record of U.S. governmental technology acquisition programs is mixed. Examples of technically successful programs include:

- The 60 successful military satellite launches since the last major failures of the FBC/AR period [Horejsi, 2009].
- All but two successful space shuttle flights.
- Construction and operation of the space station.
- Some well-known interplanetary flights.
- Some successful missile defense tests.
- Numerous other successes in air-, ground-, and water-borne systems.

These examples indicate that the established SE process can be capable of delivering successful complex systems 'when practiced properly.' However, the scope and boundaries of 'proper SE practice' are not always clear, which can lead to dramatic system failures. Theoretically, SE is supposed to consider the entire life cycle of the system being designed. Several recent and rather spectacular failures reveal fuzzy life cycle definitions and hazy boundaries between SE and other domains:

- The collision of Russian and American satellites [Broad, 2009] has sent uncountable small objects into random trajectories that now jeopardize other space missions. This raises the question whether the collisions should be blamed on the lack of sufficient international cooperation in space or on an inadequate SE program for the monitoring of space objects.

---

[1]These failures gave rise to the popular mantra that it is possible to have any two of the three (for example, faster and better, but not cheaper) of FBC, but not all three together.

- A terminally aged U.S. military satellite was destroyed by a missile, arguably in order to avoid the risk of earth contamination [Ferster, 2008], raising the question as to why the satellite disposal problem was not included in the life cycle SE of the program.

- The tragic end of the Columbia space shuttle, diagnosed by the investigation board as "the foam did it, but NASA culture allowed it" [NASA, 2003], and the misuse of O-rings in cold weather leading to the Challenger disaster [House of Representatives, 1986], provoke the question: To what degree should the organizational culture and effective SE be interrelated and inclusive of one another?

- The U.S. Coast Guard received 27 boats deemed unusable for service [O'Rourke, 2008], and the matter is now under litigation to determine fault among several disjointed parties who (mis)conducted the fragmented SE efforts.

The list of failed programs is longer and these are only examples. A comprehensive program-by-program study of the effectiveness of SE, and of the reasons for both successes and failures, remains wanting at this time, and the confounding of SE and the government acquisition system only blurs the picture.

5.2 WASTE IN TRADITIONAL SYSTEMS ENGINEERING

Waste is abundant in all programs. Failed systems represent the most dramatic manifestation of waste, but even the most technically successful programs are burdened with significant amounts of waste.

As mentioned in Chapter 1, Systems Engineering is an inherent part of Product Development, therefore, we do not differentiate between the two when talking about waste.

Waste in traditional PD programs result from a large number of causes: craft mentality of engineers (the conviction that everything is unique, requires a special solution, and cannot be framed into a process), poor planning and ad hoc execution, poor coordination and communication culture, poor corporate preparations for programs, and many others [Oppenheim, 2004]. The reader is invited to scan the enabler tables listed in Chapter 7 and look at the wastes listed under each enabler to recognize that the opportunities for creating waste are vast indeed.

The book *Lean Product and Process Development* [Ward, 2007], contains an entire chapter devoted to waste, "Seeing Waste in Product Development," where the author refers to the "Waste of Knowledge."

The present book uses Ohno's original seven wastes of manufacturing, as adapted to PD by Morgan and Liker [2006]. These were shown in Box 3.1 in Chapter 3, with examples added by the present author.

Figure 5.1 illustrates the amount of wasted effort by individuals and wasted time by work packages that have been measured in several aerospace programs [McManus, 2004]. The left-hand chart in Figure 5.1 shows that only 29 – 30% of

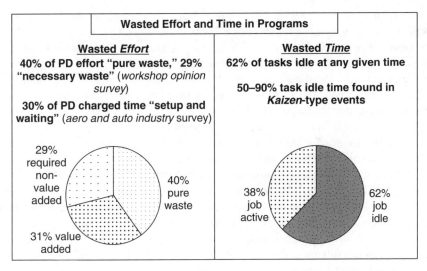

Figure 5.1 Waste Measured in Aerospace Programs [McManus, 2004].

effort adds value. The right-hand chart in the figure illustrates that a given task is being worked on for only 38% of its existence. By implication from the first chart, of that 38% time, only about a third of it, or 12% of the total task lifetime, has any value-added activity. Thus, 88% of the time represents a productivity reserve. From this author's perspective, that 88% of program time waits for Lean to unleash the reserve and make the programs so much better.

Figure 5.2 illustrates the most frequently observed waste in 27 aerospace PD programs [Slack, 1998]. It is interesting that the most frequently cited waste is waiting (people waiting for information, or information waiting for people), while

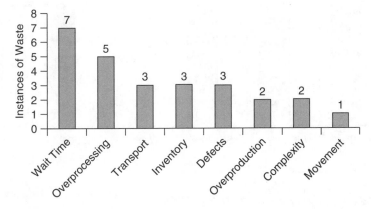

Figure 5.2 Most Frequently Observed Waste in 27 Programs [Slack, 1998].

the second most cited waste is overprocessing. To paraphrase the finding, one person is waiting, while another person is overprocessing.

Three recent reports by the Government Accountability Office (GAO) and the Department of Defense (DoD) on the condition of major programs indicate serious problems, exceeded budgets and schedules, and diminished value. Figure 5.3 [GAO, 2008b] compares the originally contracted to the recently updated cost of six major government programs. The chart shows that five of the six space programs experienced cost increases between $1 and $6 billion.

Similarly, Figure 5.4 shows that each of these five programs also encountered schedule overruns ranging from two to seven years [GAO, 2008b]. The worst case of these is the Space Based Infrared System (SBIRS). This program is roughly seven years behind schedule, $6 billion over its contracted cost, and has triggered four Nunn–McCurdy breaches (provisions that require U.S. Congress notification when a program cost increases by 15%, and termination of the program if the cost surpasses 25% of the original estimate, unless the Secretary of Defense presents documentation explaining the necessity of continuing the program for national security and no alternative lower cost option exists).

In addition to the monetary penalty, the nation's military capability and buying power are weakened when the contracted budget and schedule are exceeded. This

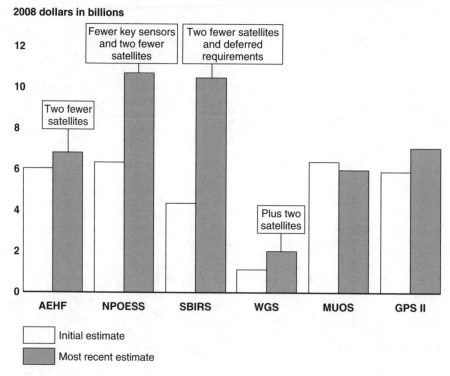

Figure 5.3   Initial and Most Recent Estimates of Total Program Costs [GAO, 2008b].

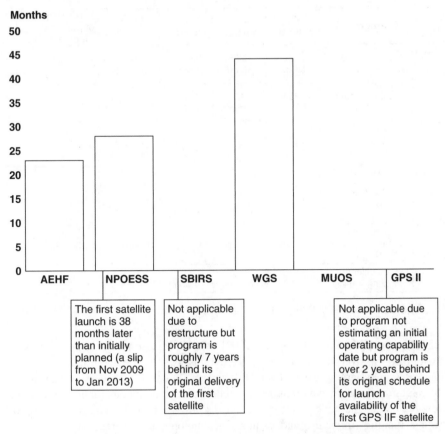

Figure 5.4 Schedule Overruns for Five Space Programs [GAO, 2008b].

is illustrated in Figure 5.5 [GAO, 2007]. The study demonstrates that the rise in cost reduces the quantity of systems delivered. As shown in Figure 5.5, the Joint Strike Fighter (JSF, now called F-35) aircraft program has gone over $25 billion beyond the initial estimate of $196.5 billion, with a loss of more than 500 aircraft for the U.S. military. (Estimates for the year 2010 have the program cost at over $340 billion, which translates to more than a $143 billion overrun and proportional reduction in the number of aircraft to be delivered.) Thus, the military attains the expected capability years after the expected date, and with a lower quantity than contracted.

The GAO [2008a] has studied the source and impact of quality problems that are the major cause for schedule and budget overruns in 11 programs (Figure. 5.6). The three main sources of program quality problems identified in Figure 5.6 are SE, manufacturing, and supplier quality. Eight of the 11 programs listed have identified poor Systems Engineering practices as a significant source of quality problems and the resultant schedule delay and exceeded cost.

Program	Initial estimate	initial quantity	Latest estimate	Latest quantity	Percentage of unit cost increase
Joint Strike Fighter	$196.5 billion	2866 aircraft	$223.3 billion	2458 aircraft	32.8
Future Combat Systems	$85.8 billion	15 systems	$131.7 billion	15 systems	54.1
V-22 Joint Services Advanced Vertical Lift Aircraft	$36.9 billion	913 aircraft	$50.0 billion	458 aircraft	170.2
Evolved Expendable Launch Vehicle	$16.0 billion	181 vehicles	$28.6 billion	138 vehicles	134.7
Space Based Infrared System High	$4.2 billion	5 satellites	$10.4 billion	3 satellites	311.6
Expeditionary Fighting Vehicle	$8.4 billion	1025 vehicles	$11.3 billion	1025 vehicles	33.7

Figure 5.5   Examples of Reduced Buying Power and Military Capability [GAO, 2007].

DoD acquisitions chief [Young, 2009] offered an assessment of the causes of cost growth in over 40 major weapons programs. Among the major factors causing increases in program cost and schedule delays, Young lists excessive and unstable requirements; runaway requirements changes; discontinuous funding; notorious *low-balling* of the budget, schedule, and risk, driven by the desire to get the programs running; immature and exquisite technology; management and execution issues; and "the DoD eyes [being] bigger than its stomach." Young did not parse the blame between the acquisition system and SE. In 2009, the U.S. DoD closed several nonperforming programs (F-22, the Future Combat Systems, Presidential Helicopter, and several smaller programs), and published several policy memoranda (e.g., [Carter, 2010]; [Kerber[2] and Vitto, 2009]) that call for significant productivity improvements in governmental PD programs. Thus, the timeliness for Lean SE could not be better, but the creators of LEfSE did not know it at the time.

Waste is abundant in all programs. Failed systems represent the most dramatic manifestation of waste, but even the most technically successful programs are burdened with significant amount of waste, which can be reduced using Lean.

---

[2]Kerber and Vitto [2009] use the word *lean* explicitly: "We envision a lean, commercially based acquisition model ..."

System	Source of quality problem			Impact of quality problem	
	Systems Engineering	Manufacturing	Supplier quality	Cost (dollars in millions)	Schedule
Advanced SEAL Delievery System	X		X	$87	program halted
Advanced Threat Infrared Countermeasure/ Common Missile Warning System	X	X		$117	5-year delay
Expeditionary Fighting Vehicle	X			$750	4-year extension to system development
F-22A	X	X	X	$400	No schedule impact to program
Global Hawk	X		X	$239	4-month production slip for sensor suite
Joint Air-to-Surface Standoff Missile		X	X	$39	Program deferred
LPD 17 Amphibious Transport Dock	X	X	X	$846	3-year delay
MH-60s Fleet Combat Support Helicopter		X		No cost impact to program	6-month production slip
Patriot Advanced Capability-3	X	X	X	$26	6-month delay
V-22 Joint Services Advanced Vertical Lift Aircraft	X			$165	Flights operations halted for 17-months
Wideband Global SATCOM		X	X	$10	18-month delay for initial operating capability

Figure 5.6 Weapon System Quality Problems and Impact [GAO, 2008a].

5.3 BEGINNINGS OF LEAN SYSTEMS ENGINEERING

The birth of *Lean Systems Engineering (LSE)* is traced to the first meeting of the LAI Educational Network in March 2003. That year, the LAI consortium invited universities to join the new LAI Educational Network (EdNet). The EdNet mission is to collaborate on the development and dissemination of Lean curricula, including incorporation of research findings. Starting with LMU, at the time of this writing the EdNet has grown to 55 universities in the United States, Brazil, China, Europe, and Mexico. The EdNet members soon organized themselves into a number of small working groups, one devoted to LSE and another to LPD, intending to develop communities of practice in these new fields. The participants shared an understanding that SE is a sound process but is not practiced as well as it could be.

Subsequently, Murman [2003] included the LSE topic in two lectures of a graduate course on Aircraft Systems Engineering. The LAI team of Rebentisch, Rhodes, and Murman [2004] laid some theoretical foundations for LSE. Additional concepts and case studies were contributed by a panel on LSE at the 2004 Annual Symposium of International Council on Systems Engineering (INCOSE) [Rhodes et al., 2004].

These early works defined the synergy of Lean and Systems Engineering as (paraphrased):

> "*Systems Engineering* which grew out of the space industry to help deliver flawless complex systems is focused on technical performance and risk management. *Lean* which grew out of Toyota to help deliver quality products at minimum cost is focused on waste minimization, short schedules, low cost, flexibility, and quality. Both have the common goal to deliver system lifecycle value to the customer. *Lean Systems Engineering* is the area of synergy of Lean and Systems Engineering with the goal to deliver the best lifecycle value for technically complex systems with minimum resources."

This synergy gave rise to the subsequent definition of LSE included in the INCOSE Handbook [2010]:

> *Lean Systems Engineering* is the application of Lean principles, practices, and tools to Systems Engineering in order to enhance the delivery of value to the system's stakeholders.
>
> —INCOSE

5.4 LEAN SYSTEMS ENGINEERING WORKING GROUP OF INCOSE

During 2004–2006, the small LAI EdNet LSE group of mostly university professors met several times and enjoyed interesting discussions; however, not much progress was made. In order to move the LSE project at a faster pace, at the end of 2005, the author made a proposal to INCOSE to form a new Lean Systems Engineering Working Group (LSE WG), hoping to draw from the collective wisdom of the large and international membership of SE practitioners who belong to that learned society. The proposal was accepted, and the LSE WG became the

39th WG of INCOSE. The first meeting (rather poorly advertised) drew 30 people, which indicates a high level of interest in the idea of applying Lean Thinking into SE. At the time of this writing (March 2011), the LSE WG has grown to 200 individuals, all unpaid volunteers, and is currently the largest Working Group of INCOSE. Most of the members are experienced industrial and governmental SEs, but the experience in Lean Thinking is less common. The individuals most experienced with Lean acted as leaders in the project of developing LEfSE (listed in the Acknowledgments).

The use of term *Lean* in the context of SE initially met with concern that this might be an attempt to "re-package the Faster-Better-Cheaper initiative," leading to cuts in SE at the time when the profession is struggling to increase the level and funding of SE effort in programs. Hopefully, this book categorically disproves these concerns. To restate: 'Lean SE' does not mean 'less SE' but 'more and better SE, leading to subsequent streamlined program execution.'

The LSE WG devoted the first 18 months to conceptual and administrative tasks (creation of public website [INCOSE LSE WG, 2008] and mailing list, definitions, recommended readings, and formulation of the Charter), as well as presentations and panels devoted to various ideas how to proceed. Box 5.1 contains the Charter of the Working Group.

Box 5.1 Charter of the INCOSE Lean System Engineering Working Group

It is our goal to strengthen the practice of Systems Engineering (SE) by exploring and capturing the synergy between traditional SE and Lean. To do this, we will apply the wisdom of Lean Thinking into SE practices integrating people, processes, and tools for the most effective delivery of value to program stakeholders; formulate A Body of Knowledge of Lean SE; propose amendments to the INCOSE SE Handbook with critical elements of Lean SE; and develop and disseminate training materials and publications on Lean SE within the INCOSE community, industry, and academia.

Since October 2007, the main effort of the Working Group has been devoted to the development of Lean Enablers for Systems Engineering, summarized in Chapter 6.

5.5 VALUE IN LEAN SYSTEMS ENGINEERING

In traditional SE, value to the customer is formulated using requirements, first the top-level or *customer requirements*, then detailed derived requirements allocated for all subsystems at all levels. The value proposition to be captured must involve not only explicit requirements and related documents, but also ***unspoken*** *requirements* defining needs, context, operations, interpretations, interoperability and compatibility characteristics, as well as a good understanding of customer culture.

The process of capturing top-level requirements is difficult to perform and is notorious for poor results. Past experience indicates that managers of many programs who are eager to get underway tend to rush through this phase without a robust process, ending in incomplete, incorrect, or conflicted requirements that burden subsequent programs with waste [Young, 2009]. Poorly formulated requirements can significantly increase program cost and lead time, and in extreme cases even torpedo entire programs (e.g., the recent presidential helicopter). Long duration of a program tends to introduce additional requirement instability due to the change in need or threat, which cannot be foreseen at the beginning of the program.

In complex government programs, value formulation is a difficult process, not only because of complex technology, but also because of unstable program funding, dissolved management, and policy and politics. The effort may easily take many years and involve thousands of stakeholders, including future system users, the government acquisition bureaucracy, contractors, suppliers, politicians, and lobbyists. Because of competing pressures, value may easily end up suboptimized, benefiting not future users, but rather a group of stakeholders who exert the strongest pull. For this reason, Lean Systems Engineering strongly promotes program optimization in the next larger context. Specifically, in complex national programs, this author believes, the proper context is the good of (the value to) the nation rather than the good of any single contractor, supplier, politician, community, or military unit.

In many technologically complex programs, military customers lack expertise to describe the needs clearly, and more often than not must be assisted in the task by value creators (the prime contractors) or a proxy organization through extensive efforts of interaction, cooperation, and clarification. In *Lean Systems Engineering* both customers and contractors have a responsibility to formulate requirements as well as the state of the art permits, without blaming one another for inadequate effort, while working together as a seamless team of honest, open, trustworthy partners who share the same goal.

Requirement stakeholders often ignore the fact that a requirement is an imperfect and inherently ambiguous means to describe need. Typically, a requirement is a sentence containing several words. Written in a natural language, especially one as rich as English, where each word in a dictionary has several meanings, it is inherently ambiguous in the linguistic sense. Additional ambiguity arises because of handoffs: The person writing the requirement has in his or her mind the rich context of the need, while the person reading the requirement sees only the requirement text. Because of these structural communicative disjunctions, it is critically important not only to make every effort to make all requirements crystal clear and complete, but also to create the means to clarify requirements without causing *requirement creep*, properly planning effective and efficient channels for clarifications.

In *Lean Systems Engineering* we continue defining value using requirements. But, in view of the above difficulties, we place a significantly higher emphasis on the quality of the effort of formulating and clarifying the requirements and make certain that all the above potential pitfalls are minimized. We also promote development of a robust process for capturing and formulating requirements.

Over the years, the number of top-level requirements in government programs grew at a fast rate, routinely reaching many hundreds and thousands in recent programs. This increasingly drove program bureaucracy and made programs costlier, longer, and usually more frustrating to all stakeholders. For comparison, it is fascinating to recall that the early Apollo program, without a doubt the most dramatically successful space program in the entire history of human civilization, started with only three requirements pronounced by President J. F. Kennedy (paraphrasing): "(1) Take man to the moon, (2) and back, (3) safely"! Similarly, the highly successful U2 aircraft program started only with a few requirements defining the flight altitude, speed, endurance, and payload [Rich and Janos, 1994]. Perhaps there is a lesson in it for Lean Systems Engineering?

In Lean Systems Engineering we define value using strong words to reflect the need for a high level of excellence, as shown in the following box:

> *Value* in Lean Systems Engineering is defined as a flawless product or mission delivered at minimum cost, in the shortest possible schedule, fully satisfying the customer and other stakeholders during a product or mission lifecycle.
>
> —Lean SE WG, INCOSE

The word *flawless* is intentionally strong. We use it to emphasize that the aim of Lean Systems Engineering is nothing short of excellence. It should be interpreted as the asymptote of excellence toward which the field of Lean Systems Engineering aims and converges by continuous improvements. It should also be regarded as a guidepost for all stakeholders. But, of course, it is not a promise that Lean Systems Engineering alone can deliver the flawless system. Besides LSE, many other accomplishments must occur in order to deliver *flawless* value: excellent SE process considering the entire system lifecycle, excellent domain engineering and testing, excellent verification and validation, excellent supply chain, excellent program management and leadership, and excellent people, processes, and tools throughout the program.

PERFECT IS THE ENEMY OF GOOD, THOUGH.

Chapter **6**

Development of Lean Enablers for Systems Engineering (LEfSE)

"Aim for the asymptote of SE excellence"
— from the strategy of Lean Systems Engineering Working Group of INCOSE

6.1 STRATEGY

The team developing the enablers started with the following two positions:

- The established SE process is regarded as technically sound if used and funded properly, but is often burdened with waste, therefore it should benefit from Lean Thinking. In other words, the team decided to try improving a practice that has a lot of strengths, but is not as good as it could be.
- Waste in programs should be regarded as a productivity reserve to be unleashed using Lean Thinking.

During the development of LEfSE, the team (the names and affiliations are listed in Acknowledgments) set up the following strategy:

1. The enablers are intended for industrial and governmental practitioners.
2. Both SE and Lean represent challenging areas for research as they are grounded in industrial and government practice rather than in laboratories, theory, or mathematics. In addition, most large programs that use SE are

Lean for Systems Engineering with Lean Enablers for Systems Engineering,
First Edition. Bohdan W. Oppenheim.
© 2011 John Wiley & Sons, Inc. Published 2011 by John Wiley & Sons, Inc.

proprietary, classified, with discontinuities in execution. Therefore, it is extremely difficult to gather explicit data and test hypotheses from such programs. SE case studies are not available in the public domain with sufficient detail to enable development of Lean practices. Where data is available, it is often of inadequate resolution or quality. Therefore, our team concluded that the development had to be based on collective wisdom and experience of the LSE WG members. Webb [2008] discusses development based on *tacit knowledge* at some depth.

3. Based on their lifetime experience, the team members were asked to identify those practices which deliver best value with minimum waste, the *dos and don'ts* of SE and the relevant aspects of enterprise management (including PD and supply chain management), applying Lean Thinking to the individual steps of the System Life Cycle Process Overview[1] [INCOSE, 2007].

4. The members were to "aim for the asymptote of SE excellence."

5. Constraints such as the present acquisition regulations and various company policies and traditions were to be disregarded. The intent was to create an ideal generic master checklist of best practices and to leave implementation to individual organizations. (Note: in spite of this strategic assumption, all Lean Enablers presented in Chapter 7 turned out to conform to the current U.S. government Acquisition System.)

6. The new enablers should not repeat information already covered in the INCOSE Handbook, which was regarded as sound but lacking Lean Thinking. The team used the [INCOSE, 2007] as the baseline, but the newer version 3.2 [INCOSE, 2010] is recommended. The 26 processes listed in version 3.2 are regarded as sound, but lacking Lean Thinking.

7. In order to reduce subjectivity, the project results were endorsed by surveys and compared with recent GAO and NASA recommendations.

6.2 DEVELOPMENT OF LEfSE

The development of LEfSE followed established design phases, including Conceptual, Alpha, Beta, Prototype, and Version 1.0. These phases are summarized in Box 6.1. Editing of candidate enablers involved tradeoffs between importance, completeness, brevity, and clarity. Being mindful of the fact that SE is not and should not be a functional island but is like a nervous system interacting with the entire PD *body*, the team added enablers dealing with critical interfaces between SE and other enterprise management areas: project management, suppliers, developmental and test engineering. Selected enablers were adapted from the following sources: Gittell [2003], LEM [1996], Leopold [2004], Morgan and Liker [2006], Oppenheim [2004], Rebentisch [2005], Rockwell Collins [2007], Stanke [2001], and others from the rich menu of the LAI and University of Michigan research products, and from individual members' experiences.

---

[1] The process chart therein is credited to ISO 15288.

Box 6.1 Development of Lean Enablers for Systems Engineering [Oppenheim et al. 2010]

DEVELOPMENT PHASE	ACTIVITIES	OUTCOME	TEAM AND NUMBER OF EXPERTS
Conceptual Design	• Brainstorming meeting to identify best SE/PD practices (other than those in SE Handbooks) based on Lean Thinking	Captured 16 pages of ideas.	Beta Team (8 individuals)
Alpha	• Numerous iterations of enabler drafts. Attempt to edit into callout boxes in INCOSE SE Handbook Input-Process-Output charts. Found impractical and changed the format to standard text, listed under eight Lean headings • Added relevant enablers from LPD literature	Alpha enablers	• Murman and Oppenheim
Beta	• Editing iterations • Designed Beta survey asking to rank enablers' Importance and Use • Beta version reviewed by LSE WG	• 160 Beta enablers. • 29 surveys returned w/comments	• Beta Team edited. • Beta Survey returned by 19 SEs from MAAC and 10 from INCOSE • 40 members of LSE WG reviewed Beta.

Prototype	• Enablers regrouped into Six Lean Principles • Rounds of negotiations and editing. • Prototype survey of Importance and Use • Comparisons with NASA and GAO studies • Decision to release online	194 Prototype enablers, including headings, organized into six Lean principles	• Prototype Team (10 individuals) • Prototype survey returned by 26 SEs at large
Version 1.0	• Cosmetic edits • Set up for formal online changes • Creation of related products (Quick Reference Guide, brochure, articles, video)	V.1.0 (194 enablers (147 subenablers + 47 headings) listed under six Lean Principles) released online	• Three LSE WG Co-Chairs
Future Continuous Improvement Process and Dissemination	• Anyone can submit change request; WG members to add arguments for and against; bi-annual voting by WG and new releases • Dissemination of LEfSE to academia, industry, government	• Formal on-online change request process, designed for voting by WG • Training charts	• 171 members of LSE WG (September 2010)

The development process has been described in detail in Oppenheim et al., [2010] and is not repeated here.

A formal online process has been set up for future improvements.

6.3   SURVEY

The final Prototype was presented to the LSE WG at the INCOSE Symposium in Utrecht, May 16–19, 2008. Several WG members decided to begin implementing the Prototype enablers in their companies without waiting for the final version.

Ideally, the enablers should be validated by comparing the performance (for example, program cost and schedule, the value delivered, and stakeholder satisfaction) between traditional programs and those following the LEfSE. This, of course, is not practical, because many recent governmental programs take years, some as long as 10 to 15 years or more. This problem is compounded by the fact that no two programs are repeatable, and also because implementing all LEfSE would be a challenge to most programs. Instead, a quick reaction from the SE practitioners was needed. Therefore, the Prototype team decided to create a survey asking the SE community at large to rank the enablers. A survey was designed listing all Prototype enablers and asking for the rankings of importance and use of each enabler. For simplicity, in the survey we used the term *enabler* to denote both enablers (headings) and subenablers (individual actionable practices). The instructions and scale used were as follows:

- Rank the importance of the given enabler to the effectiveness of SE based on your professional experience, not necessarily limited to current programs or company.[2]
- Rank the current use of the listed practice in the industry, again based on your entire experience.
- Use the scale: 2 = strongly agree [with the given enabler importance or use], 1 = agree, 0 = neutral, −1 = disagree, −2 = strongly disagree.

The Prototype survey was distributed to about 100 to 150 practitioners of Systems Engineering at large in several western countries. Twenty-six responses were completed, many with comments (i.e., the response rate was about 17–26%). Figure 6.1 summarizes the Prototype survey results. The importance in Figure 6.1 was ranked high, with most enablers ranked at an average of 2, some at 1, and none at 0 or below (paraphrasing: Very Important or Important). In contrast, the use ranking shifted by 1.3 scale units to the left (less used). It was gratifying that the respondents at large ranked the LEfSE Prototype as both important and needed overall for SE effectiveness.

---

[2]The words "not necessarily limited to current programs or company" were added to overcome the need to have the survey responses approved by strict export controls required in some defense companies.

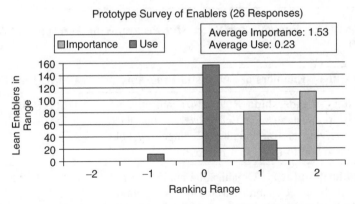

Figure 6.1 Survey Rankings of Enabler Importance and Use [Oppenheim et al. 2010].

6.4 BENCHMARKING WITH NASA AND GAO RECOMMENDATIONS

Version 1.0 of LEfSE was compared to the recommendations made in the recent studies by GAO and NASA of U.S. government programs. NASA [2007] bench-marked the practices of major aerospace companies in an attempt to capture the key enabling factors and best practices that lead to successful programs. NASA chose industry leaders with "proven outstanding achievements in producing complex systems."[3] Box 6.2 compares NASA's *key enablers* with LEfSE, and Box 6.3 compares NASA *best practices* with LEfSE. Clearly, NASA's key enablers and best practices are very *top-level*. Each NASA enabler and practice is matched by several detailed LEfSE enablers. It was gratifying to find that both NASA's *key enablers* and *best practices* are totally consistent with the LEfSE, but the LEfSE are much more comprehensive and actionable. This convergence of thinking provided further endorsement of LEfSE.

Similarly, GAO [2008b] published a summary of best practices from recent commercial space programs. Box 6.4 compares the GAO recommendations to the corresponding LEfSE. Again, the two sets are fully consistent and aligned, but, in most cases, LEfSE are more comprehensive and actionable.

6.5 VERSION 1.0 AND AWARDS

The original plan called for keeping only those Prototype enablers that the survey ranked on the importance scale at 2 or 1 (with the cutoff value 0.5) and deleting or editing those with rankings of −2, −1, or 0. As shown in Figure 6.1, all enablers passed the trigger value of 0.5. The LSE WG co-chairs decided that, after cosmetic

---

[3]Raytheon Missile Systems, Army Aviation & Missile Research and Development & Engineering Center (R&M and Software Engineering Directorates), Boeing Commercial Aircraft Division, Boeing Satellite Development Center, and Lockheed Missile & Fire Control Systems.

Box 6.2 Key Enablers for Successful Programs in Aerospace [NASA, 2007]

NASA "Key Enablers of Successful Programs"	LEfSE Enabler #
Visionary Leadership: Role of organizational leadership in establishing a clear overarching purpose, deriving and articulating a compelling but credible vision to fulfill that purpose.	1.2.6, 1.3.1, 3.4, 5.5, 5.7, 6.2
Capability Maturity: Organization attainment of high levels of "Capability Maturity" to support and facilitate the undertaking of complex systems development.	2.2, 2.3, 2.5, 3.3, 3.5, 3.6, 5.2, 5.3, 5.4, 5.6
Systems Engineering Culture:—A pervasive mental state and bias for Systems Engineering methods applied to problem solving across the development life cycle and at all levels of enterprise processes.	1.2, 1.3, 2.2.3, 2.6, 3.4, 3.6, 5.2
Design Robustness Mindset: High levels of focus on system safety and reliability driven by a bias toward achieving robustness, supported by the cultural attitude of "Failure is not an option."	2.5, 5.2, 5.3, 5.4, 5.6, 5.7, 6.3
Accountability Structure: Effective decision making accomplished through clearly defined structures of assigned responsibility and accountability for decisions at appropriate levels and phases of system development.	5.2, 6.2, 6.3

Box 6.3 Best Practices of Top Performing Aerospace Companies [NASA, 2007]

NASA "Best Practices"	LEfSE Enabler #
Leading with Vision: Sharing the Vision, Providing Goals, Direction and Visible Commitment	1.2.3, 1.2.6, 1.3.1, 3.2.6, 3.5.2, 5.5, 6.2.1, 6.2.10, 6.2.11
Focusing on Requirements: Mission Success Driven Requirements and Validation Process	1.2, 1.3, 3.2
Achieving Robust Systems: By Rigorous Analysis, Robustness of Design, HALT/HASS testing	1.2.3, 1.2.4, 2.2.3, 2.3.4, 2.4.2, 2.4.3, 3.2.5, 5.2.1a
Models and Simulation: Model-based Systems Engineering with "seamless" models, validated with Experts	1.3.3, 2.3.2, 3.2.4, 3.2.5
Visible Metrics: Effective measures, visible supporting data for better decisions at each organizational level	2.6, 3.7
Systems Management: Managing for Value and Excellence throughout the Life Cycle	1.2, 1.3, 2.2, 2.3, 2.4, 4.2, 5.2, 5.5
Building Culture: Based on Foundation "Systems" Principles, Continuous improvement	5.2, 5.6, 5.7, 6.2, 6.3

Box 6.4 Commercial Best Practices during Program Development [GAO, 2008b]

GAO Commercial Best Practices during Program Development	Lean Enabler #
• Use quantifiable data and demonstrable knowledge to make go/no-go decisions, covering critical facets of the program such as cost, schedule, technology readiness, design readiness, production readiness, and relationships with suppliers.	2.5, 2.6, 3.2, 3.3–3.7
• Do not allow development to proceed until certain thresholds are met—for example, a high proportion of engineering drawings completed or production processes under statistical control.	2.6.4, 5.2
• Empower program managers to make decisions on the direction of the program and to resolve problems and implement solutions.	1.2.5, 2.5, 3.5.7, 5.5, 6.2.8
• Hold program managers accountable for their choices.	5.5
• Require program managers to stay with a project to its end.	5.5
• Hold suppliers accountable to deliver high-quality parts for their product through such activities as regular supplier audits and performance evaluations of quality and delivery, among other things.	2.5
• Encourage program managers to share bad news, and encourage collaboration and communication.	3.5, 3.7

changes, the Prototype can be released online as Version 1.0, with the release date of January 2009 during the INCOSE annual International Workshop in San Francisco. Appendix 1 describes the web page.

While Version 1 of LEfSE represented the end of the intended development cycle, it should not be regarded as "final." Progress in the knowledge of SE and

PM, experience learned from the use of LEfSE in actual programs, and future changes in acquisition policies by different governments may require continuous improvements. The INCOSE LSE WG has implemented an online process for capturing the new experiences and suggesting changes.

As mentioned above, Oppenheim, Murman and Secor [2010] published a journal article documenting the background and development process and the text of LEfSE. These authors, together with the Lean Systems Engineering Working Group, were honored with two following prestigious awards:

- The 2009 INCOSE Product of the Year
- The 2010 Shingo Award for Research and Publication

Besides personal gratification, the awards are significant in that both *Systems Engineering* and *Lean* communities independently recognized the *Lean Enablers for Systems Engineering*.

Chapter 7

Lean Enablers for Systems Engineering

Promote excellence under "normal" circumstances instead of hero-behavior in "crisis" situations. (Lean subenabler 5.2.2)

Create a vision that draws and inspires the best people. (Lean subenabler 6.2.1)

Treat people as most valued assets, not as commodities. (Lean enabler 6.5)

7.1 ORGANIZATION

Version 1.0 of LEfSE is organized into six Lean Principles, each listing several enablers, and each enabler listing several subenablers. The six Lean Principles are: Value, Map the Value Stream, Flow, Pull, Perfection, and Respect for People, numbered 1 to 6, respectively.

There are a total of 47 enablers and 147 subenablers. Thus, the total number of enablers and subenablers is 194, and all of them were counted individually during development (which is why some documents on the INCOSE LSE WG web page refer to 194 enablers). However, the 47 enablers should be understood only as mostly non-actionable headings for the listed detailed and actionable subenablers. In other words, LEfSE includes 147 actionable subenablers organized into six Lean Principles and listed under 47 topical headings.

Lean for Systems Engineering with Lean Enablers for Systems Engineering,
First Edition. Bohdan W. Oppenheim.
© 2011 John Wiley & Sons, Inc. Published 2011 by John Wiley & Sons, Inc.

The subenablers are be identified by three-digit numbers, where:

The first digit, "x," designates the Lean Principle.
The second digit, "y," designates an enabler (heading).
The third digit, "z," designates an actionable subenabler.

Example: 1.2.5 denotes:

Lean Principle 1: Value
Enabler 2:   Establish the Value of the End Product or System to the
                       Customer.
Subenabler 5:   Do not ignore potential conflicts with other stakeholder
                       values, and seek consensus.

In the following text, we begin each Principle with a short text summarizing the listed enablers. This is followed by tables explaining the subenablers in detail.

Fifty tables present individual subenablers; 34 tables contain several closely related subenablers each. Therefore, the total number of tables is 84 rather than the 147 subenablers in the set. This organization should be useful to readers looking for closely related practices.

The numbering of the enablers and subenablers is complex enough. Therefore, we did not number the tables. In other words, the tables should be identified by the Lean Principle, enabler, and subenabler numbers.

Each table contains the following fields:

LEAN PRINCIPLE: Lists the Principle number and name, e.g., 1 Value.
ENABLER: Lists the enabler number and full text.
SUBENABLER NUMBER: Lists the subenabler number starting from 1 under each enabler and text. **AND USE RANKINGS:** The Use ranking is presented as, for example, (U = −0.25). This is the average of all survey responses for the question: "Rank the use of the given Enabler based on your professional experience, not necessarily limited to current programs or company using the following scale: 2 = strongly agree [that the given enabler is used routinely], 1 = agree, 0 = neutral, (−1) = disagree, (−2) = strongly disagree." Note: because all subenablers listed were ranked as either (paraphrasing) *important* or *very important*, the importance ranking is not listed.
VALUE PROMOTED: This line describes those aspects of a program value that can be expected to improve when the subenabler is implemented. Some improvements listed are tangible (e.g., cost) and some are not (e.g., reduction of stakeholder frustrations). Both are critical to value creation and a well-functioning program.
WASTE PREVENTED: Lists the categories of waste prevented by the subenabler. In many tables, where the subenabler affects a broad range of program performance aspects, this is listed as "all categories of waste."
EXPLANATION: A narrative description of the subenabler, explaining why the subenabler is important to program success. This box is critical to the understanding of the subenabler.
SUGGESTED IMPLEMENTATION: Describes suggested implementation methods and steps, such as training, standards, better communication.
LAGGING FACTORS: Lists factors that resist and slow implementation of the subenabler or make it more difficult.
SUGGESTED READING: A list of books, scholarly articles, and selected conference presentations relevant to the subenabler, or the source or inspiration for the subenabler. The items are recommended in the order listed.

Many of the tables listed in this Chapter are longer than one page. Horizontal dashed lines are used to denote table pagination.

In order to emphasize that we regard the traditional SE process [INCOSE, 2010] as sound, only lacking Lean Thinking, each Principle begins with the following enabler: "Follow all [applicable to the Principle] processes in the INCOSE SE

Handbook. In addition:"—and this is followed by the enablers and subenablers developed in this project. Most emphatically, all established SE process activities, including decision making, requirements capture and development, system architecting, program planning, implementation, integration and testing, validation, risk analysis, and interface analysis are fully embraced. LEfSE are not to replace the traditional SE process; they are to make it better by adding the wisdom of Lean Thinking. Since this first enabler under each Principle is self-explanatory, no table describing it is included, and the table numbers begin with two.

7.2 TABLES WITH LEAN ENABLERS FOR SYSTEMS ENGINEERING (LEfSE)

Note: The Enablers and Subenablers are listed consecutively on page 69.

7.2.1 Lean Principle 1: Value

Summary In order to assure a successful validation, the initial phase of every program must establish:

- A comprehensive, complete, unambiguous, and detailed understanding of value to the customer.
- Not only traditional requirements, but also needs, context, operations, interpretations, interoperability and compatibility characteristics, as well as good understanding of customer culture.

Many programs tend to rush through this phase without a robust process, ending in incomplete or incorrect requirements that burden the subsequent program with waste. The enablers that follow define the necessary conditions for a task to be value added, with value to be defined by the customer, while promoting development of a robust and effective process

- of capturing the complete customer value proposition,
- disseminating it among the program team,
- training and aligning the team toward this goal,
- involving the customer and other relevant stakeholders in the process,
- and doing it with sufficient breadth and depth to avoid later waste.

Three enablers are listed, as follows. The numbers in parentheses list the number of subenablers under each enabler.

Enabler 1.1 Follow All Practices for the Capture and Development of Requirements, as Found in the INCOSE Handbook.

Lean Enablers are intended to add Lean wisdom to the traditional Systems Engineering process rather than replace the process. The present enabler recommends to continue using the INCOSE SE Handbook for all traditional processes of Systems Engineering for capturing and developing requirements and to implement the enablers that follow.

Enabler 1.2 Establish the Value of the End Product or System to the Customer (6)

These enablers define what constitutes a value-added activity (task) and what does not. The key concept here is that customer (external or internal) alone defines what value is and what is not.

Enabler 1.3 Frequently Involve the Customer (4)

This enabler addresses two critical parts of successful programs: the customer-first spirit shared among all employees and involvement of the customer in the program execution. Such programs tend to have fewer conflicts and crises; better communication channels, both within a team and with the customer, are streamlined for better delivery of value to the end customer and to reach consensus easier.

The following tables describe the subenablers individually.

LEAN PRINCIPLE 1: Value

ENABLER: 2. Establish the Value of the End Product or System to the Customer.

SUBENABLER(S) AND USE RANKINGS:
1. **Define value as the outcome of an activity that satisfies at least three conditions: (U = 0.36)**
   a. **The external customer is willing to pay for Value. (U = 0.65)**
   b. **Transforms information or material or reduces uncertainty. (U = 0.57)**
   c. **Provides specified performance** *right the first time*. **(U = 0.09)**
2. **Define value-added in terms of value to the customer and his or her needs. (U = 0.50)**

VALUE PROMOTED: Excellent and complete understanding of value to the customer stakeholders.

WASTE PREVENTED: Overproduction, overprocessing and rework.

EXPLANATION: These subenablers define what constitutes value-added activity (task) and what does not. The key concept used here is that the customers (external or internal) and their needs alone define what value is and what it is not. In order for an activity to be value-added, all three conditions (a–c) must be satisfied, as follows.
- Condition (a) intends to eliminate those tasks that the customer regards (or would regard if the customer were an expert) as being of no value, or which have no internal customer. As McManus [2004, p. 108] states: "An unreleased [...] study has found that an alarming percentage of PD process outputs are not needed by downstream processes, for program knowledge capture, for meeting regulations, contractual requirements, or quality standards, or for any other purpose. They are waste."
- Condition (b) says that the activity must be doing something positive for the customer or program, and not just be a bureaucratic or self-serving task.
- Condition (c) states that if any task could be performed *right the first time* by proper planning, preparations, coordination, training, specifications, etc., it should be so executed, and if it were not because those good practices were not followed, and now must be reworked, the original attempt must be classified as waste. Subenabler 1.2.1 is not intended to eliminate trade space exploration and other legitimate iterations, which by definition cannot be done with a single pass, but as a combined activity should be planned, prepared, and coordinated to be done well once, efficiently moving through iteration loops. In this sense they also should be done *right the first time*.

Subenabler 1.2.2 expresses the key Lean concept, that the customers and their needs alone define what value is and what it is not. In this consideration, customers can be internal or external (the external could be represented by multiple stakeholders, including users, representatives, and others).

Morgan and Liker [2006, p. 37] describe the Toyota process for capturing value: "There must be a process for identifying product-specific, customer-defined value, effectively communicating that value, and developing and executing specific, aligned objectives throughout the organization from the start of the program."

SUGGESTED IMPLEMENTATION:
- Simple training in the critical importance and meaning of value.
- High expectations and mentoring from the managers involved in value capture.
- Value-stream mapping done at the beginning of the program to identify value and waste in the planning stages.
- Program kickoffs emphasizing value—clearly sharing customer expectations with a program team.

LAGGING FACTORS:
- Poor tailoring of programs, inserting tasks from previous programs that have no justification in the present program, "cut and paste" mentality planning.
- Execution of tasks disregarding value proposition.

SUGGESTED READING:
- J.M. Morgan and J.K. Liker, *The Toyota Product Development System*. New York: Productivity Press, 2006.
- E.M. Murman, T. Allen, K. Bozdogan, J. Cutcher-Gershenfeld, H. McManus, D. Nightingale, E. Rebentisch, T. Shields, F. Stahl, M. Walton, J. Warmkessel, S. Weiss, and S. Widnall, *Lean Enterprise Value: Insights from MIT's Lean Aerospace Initiative*, Palgrave, Hampshire, 2002.
- A.C. Haggerty and E.M. Murman, *Evidence of Lean Engineering in Aircraft Programs, 25th International Congress of the Aeronautical Sciences*, Hamburg, Germany, September 2006.
- H. McManus, *Product Development Value Stream Mapping Manual*, LAI, MIT, Cambridge, MA, 2004.
- D. Lempia, *Using Lean Principles and MBE in Design and Development of Avionics Equipment at Rockwell Collins, 26th International Congress of the Aeronautical Sciences,* Anchorage, AK, September 16, 2008, paper 6.7.3.
- E.M. Murman, *Lean Aerospace Engineering*, Littlewood Lecture AIAA-2008-4, January 2008. Deborah Secor, *Implementation of Lean Enablers*, LAI Plenary, Dana Point, March 23, 2010.

LEAN PRINCIPLE 1: Value

ENABLER: 2. Establish the Value of the End Product or System to the Customer.

SUBENABLER(S) AND USE RANKINGS:
3. Develop a robust process to capture, develop, and disseminate customer value with extreme clarity. (U = 0.00)

VALUE PROMOTED: All aspects of value are affected by this subenabler: Mission or product assurance, cost, and schedule.

WASTE PREVENTED: Poorly captured value will introduce delays and added costs to a program and may even destroy it. All categories of waste are affected.

EXPLANATION: The fundamental phase of every program is requirements capture and development. Requirements should capture a comprehensive, unambiguous, and detailed understanding of value of an end product or mission to customer stakeholders, including not only the traditional requirements, but also the *bigger picture*: the need, context, operations, scenarios, customer culture, and interpretations. This subenabler promotes the development of a comprehensive and robust process executing this program phase perfectly. As discussed in Section 5.5, value capture in complex government programs is a big challenge, involving numerous stakeholders and difficulties. These reasons make this subenabler critically important: The process must work robustly and effectively, or significant waste will occur in the subsequent program. Nothing short of perfection should be expected of all stakeholders, which requires solid training, experience, knowledge, teamwork, and coordination.
Few organizations do it right (the use ranking is only 0.0). Many programs tend to rush through this phase, ending in incomplete or incorrect requirements that burden the subsequent program with waste. Requirements changes are among the most costly aspects of program execution and should be minimized by all means.

SUGGESTED IMPLEMENTATION: Since a robust process for requirements capture and development applies and is critical to all programs, it should be developed at the enterprise level, to serve all future programs. If the enterprise level process is not practical, it should be developed and implemented at the program level. The most experienced Systems Engineers in the enterprise should contribute their wisdom and experience to develop the process. Ensure those Systems Engineers truly understand the Lean principles as they develop this process. Use after-action reviews and lessons learned from previous programs before starting a new project. Perhaps consultants knowledgeable of the lessons learned should aid the effort.

LAGGING FACTORS:
- Poor formulation of customer requirements prior to the Request for Proposal.
- Lack of enterprise-level, long-term planning.
- The programs that rush to the *earned-value* phase prematurely without proper preparations and planning.
- The *cost-plus* incentives for requirements changes.

SUGGESTED READING:
- J.M. Morgan and J.K. Liker, *The Toyota Product Development System*. New York: Productivity Press, 2006.
- A.C. Haggerty and E.M. Murman, *Evidence of Lean Engineering in Aircraft Programs, 25th International Congress of the Aeronautical Sciences*, Hamburg, Germany, September 2006.
- E.M. Murman, *Lean Aerospace Engineering*, Littlewood Lecture AIAA-2008-4, January 2008.
- D. Lempia, *Using Lean Principles and MBE in Design and Development of Avionics Equipment at Rockwell Collins, 26th International Congress of the Aeronautical Sciences,* Anchorage, AK, September 16, 2008, paper 6.7.3.
- A.C. Haggerty, *The F/A-18E/F Super Hornet as a Case Study in "Value Based" Systems Engineering*, INCOSE 2004, Toulouse, France, June 20–24, 2004.

LEAN PRINCIPLE 1: Value

ENABLER: 2. Establish the Value of the End Product or System to the Customer.

SUBENABLER(S) AND USE RANKINGS:
4. Develop an agile process to anticipate, accommodate and communicate changing customer requirements. (U = 0.28)

VALUE PROMOTED: Mission or product assurance, robust budget and schedule.

WASTE PREVENTED: Rework, delays.

EXPLANATION: While every effort should be placed on capturing and developing original requirements right, both on the part of the customer/user and the contractor, experience indicates that some changes in requirements occur in almost all complex programs. The changes are caused by modified functionalities of the required system, changes in threat, accommodation of objective limitations discovered during program execution, or discovery of vastly better solutions. Given the high probability of requirements changes, a healthy PD enterprise will be well prepared for changes with agile (flexible) process. The process should capture and accommodate the changes with a minimum of disruption and rework. All changes should be properly communicated to all stakeholders to minimize information churning and waste of rework.
This subenabler should not serve as an excuse to perform sloppy capture of the value proposition and requirements at the program beginning.

SUGGESTED IMPLEMENTATION: Since an agile process for capturing and accommodating requirements changes applies to and is critical to all programs, it should be developed at the corporate level to serve all future programs. If the enterprise-level process is not practical, it should be developed at the program level. The most experienced Systems Engineers, with comprehensive understanding of Lean, should contribute their wisdom and experience to the process formulation. Emphasis should be placed on process agility (flexibility) to handle any change with minimum disruption to the program. Planning for good communications at the program level is essential to ensure that all relevant stakeholders are involved in the requirements change process at the right time in the development.

LAGGING FACTORS:
- *Cost-plus* incentives for complex requirements changes.
- Lack of enterprise-level, long-term planning.
- The programs that rush to enter the high *earned-value* phase prematurely without proper preparation and planning.

SUGGESTED READING:

- J.M. Morgan and J.K. Liker, *The Toyota Product Development System*. New York: Productivity Press, 2006.

- A.C. Haggerty and E.M. Murman, *Evidence of Lean Engineering in Aircraft Programs, 25th International Congress of the Aeronautical Sciences*, Hamburg, Germany, September 2006.

- E.M. Murman, *Lean Aerospace Engineering*, Littlewood Lecture AIAA-2008-4, January 2008.

- D. Lempia, *Using Lean Principles and MBE in Design and Development of Avionics Equipment at Rockwell Collins, 26th International Congress of the Aeronautical Sciences,* Anchorage, AK, September 16, 2008, paper 6.7.3.

- A.C. Haggerty, *The F/A-18E/F Super Hornet as a Case Study in "Value Based" Systems Engineering*, INCOSE 2004, Toulouse, France, June 20–24, 2004.

LEAN PRINCIPLE 1: Value

ENABLER: 2. Establish the Value of the End Product or System to the Customer.

SUBENABLER(S) AND USE RANKINGS:
5. Do not ignore potential conflicts with other stakeholder values, and seek consensus. (U = 0.28)

VALUE PROMOTED: Lean flow of work without stopping, backflow, or rework.

WASTE PREVENTED: Rework, overprocessing, inventory, transportation, delays.

EXPLANATION: In a typical complex program, the number of stakeholders involved is large, representing the customer community, the prime contractor, numerous suppliers, lobbyists, politicians, bankers, and others. This many stakeholders create significant potential for conflicts. Good managers will anticipate typical potential conflicts and implement measures to prevent them and seek consensus. See the "Suggested Implementation" section. The role of a Chief Systems Engineer (or analogous position) is critical. This person should be a strong consensus builder and have significant expertise and experience with both domain and stakeholders.

SUGGESTED IMPLEMENTATION: The Lean Thinking methods for preventing conflicts and seeking consensus involve leadership; open, honest, partnering and participatory management; teamwork; proactive negotiations; excellent communication skills; opportunities for stakeholders to brainstorm; frequent *integrative reviews* (e.g., weekly) where all issues are addressed and resolved before they become crises; and training and mentoring. This kind of enterprise culture cannot be achieved overnight. It must be built patiently using the best industrial examples.

LAGGING FACTORS:
- Authoritarian management style.
- Program management dissolved among many individuals.
- Rigid positions.
- Lack of good examples.

SUGGESTED READING:

- J.M. Morgan and J.K. Liker, *The Toyota Product Development System*. New York: Productivity Press, 2006.

- D. Lempia, *Using Lean Principles and MBE in Design and Development of Avionics Equipment at Rockwell Collins*, 26th *International Congress of the Aeronautical Sciences,* Anchorage, AK, September 16, 2008, paper 6.7.3.

- C. Haggerty and E.M. Murman, *Evidence of Lean Engineering in Aircraft Programs, 25th International Congress of the Aeronautical Sciences*, Hamburg, Germany, September 2006.

- A.C. Haggerty, *The F/A-18E/F Super Hornet as a Case Study in "Value Based" Systems Engineering*, INCOSE 2004, Toulouse, France, June 20–24, 2004.

- E.M. Murman, *Lean Aerospace Engineering*, Littlewood Lecture AIAA-2008-4, January 2008.

LEAN PRINCIPLE 1: Value

ENABLER: 2. Establish the Value of the End Product or System to the Customer.

SUBENABLER(S) AND USE RANKINGS:
6. Explain customer culture to program employees, i.e., the value system, approach, attitude, expectations, and issues. (U = −0.52)

VALUE PROMOTED: Better understanding and realization of value in the program by all employees.

WASTE PREVENTED: Rework, suboptimum solutions to customer needs, delays.

EXPLANATION: Establishing the value of the end product or system to a customer involves not only formal requirements, but also good understanding of a customer culture, including the value system, approach, attitude, expectations, and issues. Too often, employees working on a program are given only a myopic view of it, seeing only their assignment while lacking awareness of the *bigger picture*, customer environment,and culture. This lack of awareness produces solutions that are less than optimal for the customer, even if nominally meeting minimum requirements. It is critical that managers align all employees for the good of the customer, and that requires both solid understanding of customer environment and culture, and customer-focused culture in the contractor enterprise.

SUGGESTED IMPLEMENTATION: One of the easiest subenablers to implement: The Program Manager should summarize these points concisely, publish them, and display them in the spaces visible by all employees and other stakeholders. An ideal time to share this comprehensive view of the customer, including both challenges and opportunities, is during the program planning kickoff with stakeholders. If the only person who understands customer culture is the program manager, the product development team will never have the opportunity to be truly successful in meeting customer expectations. Computer savers can also be practiced. Periodically, all program stakeholders should be reminded of the points in meetings. All new hires should receive a briefing or a written memo on these points.

LAGGING FACTORS:
- Managers who insist that formal requirements are to be the only means of capturing value: "We do only what is required."
- Managers who see such alignment of employees as a waste of program resources.
- Preference given to "automated" management rather than leadership.
- The culture of isolated program employees communicating mostly by computers.

SUGGESTED READING:

- J.M. Morgan and J.K. Liker, *The Toyota Product Development System*. New York: Productivity Press, 2006.

- D. Lempia, *Using Lean Principles and MBE in Design and Development of Avionics Equipment at Rockwell Collins*, 26th *International Congress of the Aeronautical Sciences,* Anchorage, AK, September 16, 2008, paper 6.7.3.

- A.C. Haggerty and E.M. Murman, *Evidence of Lean Engineering in Aircraft Programs, 25th International Congress of the Aeronautical Sciences*, Hamburg, Germany, September 2006.

- A.C. Haggerty, *The F/A-18E/F Super Hornet as a Case Study in "Value Based" Systems Engineering*, INCOSE 2004, Toulouse, France, June 20–24, 2004.

- E.M. Murman, *Lean Aerospace Engineering*, Littlewood Lecture AIAA-2008-4, January 2008.

LEAN PRINCIPLE 1: Value

ENABLER: 3. Frequently Involve the Customer.

SUBENABLER(S) AND USE RANKINGS:
1. Everyone involved in the program must have a customer-first spirit.
   (U = 0.56)

VALUE PROMOTED: Employees aligned for value. Fewer frustrations among the stakeholders.

WASTE PREVENTED: All categories of waste are affected. Frustrations and crises are reduced.

EXPLANATION: Complex engineering programs offer tough challenges for everyone involved. In a moment of frustration, it is not unusual to accept a defensive attitude about oneself, one's department or team, and to lose sight of the fact that the program and the enterprise exist in order to satisfy customers. As Henry Ford famously stated: "The enterprise is not paying the payroll; it only passes the money; it is the customer who pays the payroll." Thus, a customer-first spirit and alignment of all employees and other stakeholders toward these goals are a critical part of successful programs. Such programs tend to have the least amount of bickering and crises, are streamlined for best delivery of value to customers, and reach consensus easier.

SUGGESTED IMPLEMENTATION: Ideally, at the beginning of the program, a program leader should formulate a strong customer-first spirit for all stakeholders, then disseminate it to all using the most effective means (video, memo, intranet talk, etc.). It is important to send this strong message convincingly, without patronizing or cheap sloganeering. The message should be periodically repeated and modified when necessary. New hires must receive a short training about the message. Encourage the team to ask questions for clarifications about customer requirements. If the team works together to ensure customer value, they may come up with innovative solutions not thought of by the program manager. Communication throughout the PD team is critical to ensure clear understanding of the customer value.

LAGGING FACTORS:
- *Hands-off* attitude toward the customer.
- Lack of leadership.
- Narrow technical assignments without seeing the *big picture*.
- The culture of unresolved conflicts among program stakeholders.

SUGGESTED READING:

- J.M. Morgan and J.K. Liker, *The Toyota Product Development System*. New York: Productivity Press, 2006.

- A.C. Haggerty and E.M. Murman, *Evidence of Lean Engineering in Aircraft Programs, 25th International Congress of the Aeronautical Sciences*, Hamburg, Germany, September 2006.

- R. Leopold, *The Iridium Story: An Engineer's Eclectic Journey*, Minta Martin Lecture, MIT Department of Aeronautics and Astronautics, April 23, 2004.

LEAN PRINCIPLE 1: Value

ENABLER: 3. Frequently Involve the Customer.

SUBENABLER(S) AND USE RANKINGS:
2. **Establish frequent and effective interaction with internal and external customers**. (U = 0.56)

VALUE PROMOTED: Good capture and continual understanding of the program value proposition. Ability to clarify requirements and needs when needed without introducing delays and rework.

WASTE PREVENTED: Primarily delays and rework, also all other categories of waste, including program failure.

EXPLANATION: Modern programs involve a large number of sites and geographical distribution of stakeholders. Value capture and work planning and coordination are executed using impersonal tools, with poor knowledge of who the internal customer for each task is, what are his or her particular needs, how the task contributes to the overall value, and how to optimize tasks for best value with minimum waste. This disjointed architecture is not conducive to effective communications. Therefore, it is important to organize the program so that frequent and effective communications with internal and external customers are strongly promoted and practiced from the beginning.

SUGGESTED IMPLEMENTATION: A good leader should make it clear to all program stakeholders at the beginning of a program that frequent and effective interactions with internal and external customers are expected, regardless of geographical distribution and organizational discontinuity. The leader, or the deputy, should explain in a memorandum or video how critical are good interactions for effective flow of work. The leader should personally guide the interactions among stakeholders during the value-capture phase, including trade space exploration efforts, brainstorming, negotiations, and reaching consensus. The leader should also instruct program employees to learn who the internal customer is for each task and how to interact with that person effectively before and during task execution so that it can be executed *right the first time*.

LAGGING FACTORS:
- Excessive automation of interactions among program employees.
- Lack of timely cadence reviews with the team.
- Working on a small project elements, in isolation from the rest of the team, and without understanding the *big picture*.

SUGGESTED READING:

- J.M. Morgan and J.K. Liker, *The Toyota Product Development System*. New York: Productivity Press, 2006.
- A.C. Haggerty and E.M. Murman, *Evidence of Lean Engineering in Aircraft Programs, 25th International Congress of the Aeronautical Sciences*, Hamburg, Germany, September 2006.
- R. Leopold, *The Iridium Story: An Engineer's Eclectic Journey*, Minta Martin Lecture, MIT Department of Aeronautics and Astronautics, April 23, 2004.

LEAN PRINCIPLE 1: Value

ENABLER: 3. Frequently Involve the Customer.

SUBENABLER(S) AND USE RANKINGS:
3. Pursue an architecture that captures customer requirements clearly and can be adaptive to changes. (U = 0.36)

VALUE PROMOTED: Flexible and adaptable system architecture is conducive to subsequent optimization, clear value capture, and subsequent optimization of the system.

WASTE PREVENTED: Waiting, suboptimal realization of value.

EXPLANATION: As emphasized in several other enablers, every reasonable effort should be devoted to capturing a value proposition right, including all aspects of need, operations, environment, scenarios, and customer culture. If not executed properly, value formulation will be suboptimal or incorrect, risking burdening the subsequent program with waste and potential failure. Subenabler 1.3.3 addresses the need for a system architecture that is able to capture all applicable information clearly and be adaptive to changes, permitting quick evaluations of competing ideas, architectures, and scenarios, until optimum value is formulated.
Note: Subenabler 3.2.4 below elaborates on the types of architectural items to pursue (e.g., CAE and solid models, software prototypes). The present subenabler (1.3.3) should be understood as a call for solid preparations to be planned and executed at the start of programs so that the architectural tools, checklists, and trained team members are in place when needed.

SUGGESTED IMPLEMENTATION: At the start of each program, evaluate carefully what architectural tools and checklists might be needed, develop the needed ones, and immediately begin implementing them so that they become available when needed without the waste of waiting. Consider personnel training requirements and communication paths for requirements flow.

LAGGING FACTORS:
- Architectural tools not used.
- Tendency to focus on a point design before exploring candidate architectures.

SUGGESTED READING:
- J.M. Morgan and J.K. Liker, *The Toyota Product Development System*. New York: Productivity Press, 2006.
- A.C. Haggerty and E.M. Murman, *Evidence of Lean Engineering in Aircraft Programs, 25th International Congress of the Aeronautical Sciences*, Hamburg, Germany, September 2006.
- R. Leopold, *The Iridium Story: An Engineer's Eclectic Journey*, Minta Martin Lecture, MIT Department of Aeronautics and Astronautics, April 23, 2004.

LEAN PRINCIPLE 1: Value

ENABLER: 3. Frequently Involve the Customer.

SUBENABLER(S) AND USE RANKINGS:
**4. Establish a plan that delineates the artifacts and interactions that
provide the best means for drawing out customer requirements.**
(U = 0.39)

VALUE PROMOTED: Efficiency of the capture of program value.

WASTE PREVENTED: All categories of waste, including program failure.

EXPLANATION: As emphasized in several other enablers, every reasonable
effort should be devoted to capturing value proposition right, including all
aspects of need, operations, environment, scenarios, and culture into formal
requirements. Usually, passive acceptance of customer requirements is
insufficient: Customers may lack the expertise to formulate requirements
correctly and completely. Prudent Systems Engineers should know this and
for the good of the program develop a special plan for drawing customer
requirements. The plan should define both quality artifacts and best human
and electronic interactions between program stakeholders, including the end
customer.

SUGGESTED IMPLEMENTATION:
- The Chief Engineer (or deputy) should develop a specific and robust plan
defining which artifacts to use and how, using what tools, when, and by
whom; and which interactions to set up: in person, meetings, telecoms,
emails—how often or when, who, how, where—for drawing out customer
requirements robustly and comprehensively.
- Training of the program stakeholders in the plan, inviting customer
representatives to training sessions, if practical.
- Ensure spoken and customer *unspoken requirements* are captured.
Assumptions and unclear understanding either by the customer or the
development engineers can create waste. Domain knowledge, customer
affinity, and understanding play a big role in this subenabler.

LAGGING FACTORS:
- Passive, bureaucratic and uncritical acceptance of customer-provided
requirements as final.
- Excessive artifact bureaucracy.
- Automation of interactions among program employees at the expense of
human interactions.

SUGGESTED READING:

- J.M. Morgan and J.K. Liker, *The Toyota Product Development System*. New York: Productivity Press, 2006.

- D. Lempia, *Using Lean Principles and MBE in Design and Development of Avionics Equipment at Rockwell Collins*, 26th *International Congress of the Aeronautical Sciences,* Anchorage, AK, September 16, 2008.

- A.C. Haggerty and E.M. Murman, *Evidence of Lean Engineering in Aircraft Programs, 25th International Congress of the Aeronautical Sciences*, Hamburg, Germany, September 2006.

- R. Leopold, *The Iridium Story: An Engineer's Eclectic Journey*, Minta Martin Lecture, MIT Department of Aeronautics and Astronautics, April 23, 2004.

7.2.2   Lean Principle 2: Map the Value Stream (Plan the Program)

Summary   The second Lean Principle "Map the Value Stream (often abbreviated in literature as Value Stream Mapping, VSM)" addresses program planning. It includes both formal VSM activities, as well as less formal planning aspects such as: tailoring and sequencing tasks, connecting and training stakeholders, preparing checklists, planning of communication and coordination means, planning effective metrics.

Poor preparation and planning create an immense potential for waste. To address this, the present Principle promotes both excellence of program preparations (preparation of people, tools, and processes) and excellence of program planning (establishing the most streamlined and concurrent sequence and scope of tasks and actions), including:

- Comprehensive timely planning.
- Checklist for planning of all end-to-end linked streamlined processes necessary to realize value without waste.
- Integration of the planning of Systems Engineering, project management, supply chain, and other relevant enterprise activities to avoid the frequent waste that occurs at the functional interfaces.
- The benefits of old-fashioned co-location
- The use of most experienced individuals early: during the critical planning and conceptual phases.
- Planning for maximum frontloading.
- Planning of means of coordination and communication.
- Preventing potential conflicts.
- Planning effective metrics.
- Tailoring and planning of task precedence and content for smooth flow.

Six enablers are listed under this Principle, as follows. The numbers in parentheses list the number of subenablers under each enabler.

Enabler 2.1. Plan the Program according to the INCOSE Handbook Process.

The first enabler listed under each principle emphasizes the same fundamental point: Add the wisdom of Lean Thinking to the traditional Systems Engineering process rather than replace the process. So, again, the present enabler recommends use of the INCOSE SE Handbook (or any other equivalent document), and in addition, to implement the subenablers that follow.

Enabler 2.2 Map the SE and PD Value Streams and Eliminate Non-Value-Added Elements (13)

The 13 subenablers promote:

- Excellent preparations, preferably at the corporate level, of people, processes, and tools that are common to all programs.
- Excellent, detailed, and tailored planning, including VSM of program tasks to be performed at the program beginning. Planning must involve comprehensive streamlining and elimination of waste.

Enabler 2.3 Plan for Frontloading the Program (4)

It is important to differentiate between preparations (for all programs), planning of a specific program, and program execution. The subenablers listed here promote effective planning for subsequent front-loading programs.

The early conceptual phase of a program, often perceived as fuzzy or chaotic, has a major impact on overall program success, schedule, and cost. Usually, the worst approach is to rush into a point solution prematurely. The four subenablers listed here promote the Lean Thinking approach of careful, comprehensive trade space exploration during early program phases and before converging on a point solution. This approach has been shown to yield tremendous savings during subsequent parts of programs.

Enabler 2.4 Plan to Develop only What Needs Developing (5)

The five subenablers promote the Lean approach to modularity, reusing and sharing of program assets: platforms, standards, busses, and modules of knowledge, hardware and software, as well as maximizing opportunities for future hardware and software upgrades.

Enabler 2.5 Plan to Prevent Potential Conflicts with Suppliers (4)

These four subenablers describe the Lean supply chain management and optimum relations with suppliers based on maximally seamless boundaries, long-term partnership, mutual benefits, and cultural compatibility. Such relations help prevent potential conflicts with suppliers and avoid significant waste.

Enabler 2.6 Plan Leading Indicators and Metrics to Manage the Program (5)

Recent programs are notorious for numerous bureaucratic metrics, which consume resources and fail to truthfully monitor program health and progress. Many present deceptive illusions, indicating good progress when the program is actually stalled or failing. The five subenablers listed here promote the use of leading indicators (which forecast future program health based on past performance) and the best metrics to use.

The following tables describe the Principle 2 subenablers individually.

LEAN PRINCIPLE 2: Map the Value Stream (Plan the Program)

ENABLER: 2. Map the SE and PD Value Streams and Eliminate Non-Value-Added Elements.

SUBENABLER(S) AND USE RANKINGS:
1. **Develop and execute clear communication plan that covers the entire value stream and stakeholders. (U = −0.29)**

VALUE PROMOTED: Efficient lean flow of work with minimum stopping, backflow, or rework.

WASTE PREVENTED: All categories of waste.

EXPLANATION: Effective communications are critical for program success, yet the Use ranking (U = −0.29) indicates that they are poorly practiced. Programs waste significant portions of their budgets and schedules because of inadequate communications. In extreme cases, mission failure may result from bad communications (e.g., the inconsistent units on the failed Mars Climate Orbiter mission). "Poor" can mean both "not enough good ones" and "too many bad ones," as follows.

Best communications involve the widest bandwidth: person-to-person is superior to video and phone, which are superior to email, which is superior to automated software tools. Modern trends are to execute programs over a large number of sites with geographically distributed stakeholders. This is not conducive to effective communications. It is difficult to perform good and detailed program planning (mapping both SE and PD value streams) without good knowledge of who is the internal customer for each task, how the task contributes to the overall value, and how to optimize tasks for the best value with minimum waste. As a result, plans tend to be superficial, impersonal, and, therefore, ineffective.

Waste also occurs when there are too many bad communications, e.g., when emails are abused by excessive dissemination and unneeded attachments. All employees know the phenomenon of receiving unsolicited and unneeded emails from colleagues. Yet, all received emails must be read and sorted to be saved, deleted, or acted upon. Arguably, most of us waste about one hour per day on unsolicited and unneeded email. Assuming this is the average, it constitutes 12.5% of nominal charged time. (On a $1 billion program, this waste alone costs the program $125,000,000.) The waste can be dramatically reduced with a single instruction from the Chief Engineer (or equivalent role): "Do not cc: unless: (followed by a specific instruction how to disseminate emails)."

All together, it is critical to organize from the beginning for frequent, effective, and waste-free communications. A solid plan for communications, with training, can vastly improve the quality of programs.

SUGGESTED IMPLEMENTATION: The Chief Engineer (or equivalent role) should develop an effective plan for communications among all program stakeholders and make sure that all stakeholders understand the plan. The plan should include instructions for:

- Periodic *integrative events* for addressing issues (see also Section 4.2).
- Efficient and effective meetings with specific purpose, agenda, time, intent (PATI).
- Short ad-hoc stand-up meetings
- Milestone meetings and programmatic reviews.
- Communications between program levels (between different IPTs).
- Meeting means (whether in person, by videoconference, computer, phone, etc.).
- Instructions for email (e.g., an instruction to summarize any issue on standard A3 form and not send other attachments unless requested).
- Promoting a culture of immediately asking questions rather than sitting on issue and waiting for the next *integrative event*.
- Providing the means to escalate issues, if needed.
- Mandate to identify an internal customer and coordinate task with him or her prior to executing any non-routine task
- The dos and don'ts of communications with external stakeholders (see subenabler 3.5.9).
- And any other applicable aspect of communications.

LAGGING FACTORS:
- The culture of ad-hoc unplanned communications.
- "We have worked here for years so we know what to do."
- Defensive culture of sending all documents and emails to everyone, just in case.
- Lack of knowledge on how successful companies organize effective communications.

SUGGESTED READING:
- B.W. Oppenheim, Lean product development flow, *Journal of SE*, 2004, **7**, 4, pp. 352–376.
- J.M. Morgan and J.K. Liker, *The Toyota Product Development System*. New York: Productivity Press, 2006.
- D. Lempia, *Using Lean Principles and MBE in Design and Development of Avionics Equipment at Rockwell Collins*, 26th *International Congress of the Aeronautical Sciences,* Anchorage, AK, September 16, 2008, paper 6.7.3.

LEAN PRINCIPLE 2: Map the Value Stream (Plan the Program)

ENABLER: 2. Map the SE and PD Value Streams and Eliminate Non-Value-Added Elements.

SUBENABLER(S) AND USE RANKINGS:

2. **Have cross functional stakeholders work together to build the agreed value stream.** (U = −0.04)

3. **Create a plan where both Systems Engineering and other Product Development activities are appropriately integrated.** (U = 0.30)

4. **Maximize co-location opportunities for SE and [other] PD planning.** (U = 0.17)

VALUE PROMOTED: Good detailed planning is conducive to technical excellence, and on-budget, on-schedule execution. It is conducive to synchronization of activities, reduction of errors/defects, and better cooperation across divisions and suppliers. Good planning also highlights coordination opportunities across functional groups and better stakeholder understanding of program lifecycles.

WASTE PREVENTED: Poor planning contributes to all categories of waste, including program failure.

EXPLANATION: These three subenablers promote excellent program preparations and planning. Most successful programs indicate that best planning is done by small co-located teams of domain experts and resource managers who brainstorm and negotiate together for a few days or weeks until they create a detailed plan based on consensus. Planning teams should be led by experienced Chief Engineer (or equivalent role) and should be comprised of domain experts, technically competent and team-focused managers of major functions and major suppliers, and, where necessary, representatives of user, customer, and corporation, plus a few staff assistants. Simple arithmetic indicates that the cost of co-location (travel costs) for a team of 10–15 such individuals for a few weeks may be on the order of $50,000–$100,000, vastly cheaper than the typical consequences of bad planning, including program failure. Co-location of the critical team is conducive to detailed value-stream mapping, including the planning of IPT levels, major resources, gates and milestones, *integrative events*, and the critically important frontloading and trade space exploration, utilizing best experiences, and lessons learned. Such planning can identify risks, actions, opportunities, and gaps early and permit early mitigation actions. It is important to integrate the planning of SE, IPTs, program management, functional engineering, suppliers, and other relevant enterprise activities to avoid the frequent waste that occurs at the functional interfaces. Ideally, each planning team should arrive at a consensus that all team members will be able to support with resources. It is critical that planning include not only SE activities but also all other critical PD activities, well integrated. Planning should be held in one large *war room*, with plan drafts and final version shown on walls, visible for all to see. (See Section 4.2.)

As described by Morgan and Liker [2006, p. 145], Toyota successfully integrates functions and programs because:
1. The customer is first.
2. The Chief Engineer is revered.
3. The Chief Engineer has the power of executive sponsorship.
4. General Managers understand the importance of serving customers and cross-functional cooperation.
5. Junior employees respect senior employees.

SUGGESTED IMPLEMENTATION: The implementation steps should include, in chronological order:
1. Nominate a Chief Engineer (CE, or equivalent role) for each program. The person must be experienced in the domain, in Systems Engineering, in Lean, and be a good trusted leader.
2. Let the CE organize the planning phase.
3. Allow the best team of stakeholders to co-locate.
4. Map the high-level product development value stream for the program, at the kickoff, with the key stakeholders present.
5. Do not quit the mapping until a detailed, comprehensive, experience-based and consensus-based value-stream map is completed for the program, or at least for the first few major program milestones, if it is impossible to plan the entire program.
6. Identify dependencies and precedence in the value stream (predecessor task, successor task, etc.) to ensure the stakeholders understand their internal sequencing and connectivity.
7. Provide a *big picture* of the program to stakeholders by using the value-stream map.

LAGGING FACTORS:
- Lack of a Chief Engineer (or equivalent role) who has solid experience in the domain, SE, Lean, and is a good leader.
- Bureaucratic approach to SE, focused on production of artifacts rather than on engineering a great system.
- Refusal to co-locate and relying on impersonal electronic means of communications.

SUGGESTED READING:

- J.M. Morgan, J.K. Liker, *The Toyota Product Development System*. New York: Productivity Press, 2006.

- B.R. Rich and L. Janos, *Skunk Works: A Personal Memoir of My Years at Lockheed*, Little, Brown and Company, Canada, 1994.

- B.W. Oppenheim, Lean product development flow, *Journal of SE*, 2004, **7**, 4, pp. 352–376.

- A.C. Haggerty and E.M. Murman, *Evidence of Lean Engineering in Aircraft Programs, 25th International Congress of the Aeronautical Sciences*, Hamburg, Germany, September 2006.

- A.C. Haggerty, *The F/A-18E/F Super Hornet as a Case Study in "Value Based" Systems Engineering*, INCOSE 2004, Toulouse, France, June 20–24, 2004.

- E.M. Murman, *Lean Aerospace Engineering*, Littlewood Lecture AIAA-2008-4, January 2008.

- R. Leopold, *The Iridium Story: An Engineer's Eclectic Journey*, Minta Martin Lecture, MIT Department of Aeronautics and Astronautics, April 23, 2004.

LEAN PRINCIPLE 2: Map the Value Stream (Plan the Program)

ENABLER: **2. Map the SE and PD Value Streams and Eliminate Non-Value-Added Elements**.

SUBENABLER(S) AND USE RANKINGS:
5. **Use formal value stream mapping methods to identify and eliminate SE and other PD waste and to tailor and scale tasks.** ($U = -0.67$)
6. **Scrutinize every step to ensure it adds value, and plan nothing because "it has always been done."** ($U = -0.54$)
7. **Carefully plan for precedence of both SE and PD tasks (which task to feed which other tasks, with what data, and when), understanding task dependencies and parent-child relationships.** ($U = 0.42$)

VALUE PROMOTED: Excellent planning is conducive to excellent program execution.

WASTE PREVENTED: All categories of waste result from inadequate planning.

EXPLANATION: Value-stream mapping (VSM) for Product Development (PD) is a relatively new body of knowledge [see McManus, 2004]. Its power is in its ability to simultaneously identify and remove waste as well as promote program value during the planning. Basic VSM involves two phases: Mapping the Current State and Future State (Section 4.2 presents a special case where one more phase is needed). The negative use rankings for subenablers 2.2.5 and 2.2.6 above indicate that this type of planning is not much practiced yet.

The Current-State Map is a detailed process map of the program as it would be executed using traditional methods. It should show all tasks with their precedence, flow network, and control points. It is, furthermore, a starting point for subsequent identification of waste. The final VSM used on the most recent legacy program is an efficient way to begin the Current-State Map. Knowledge of a legacy program, its problems, solutions, wastes, technical approaches, and any other issues is invaluable in this step. (Corporate incentives should be introduced to collect and preserve value-stream maps from past programs.) Starting with the legacy knowledge is helpful even if only partial information is available, such as the process map or Gantt chart of an actually executed program. Participation of a high-level manager from a legacy program in VS mapping of a current program is highly helpful. Brainstorming, negotiations, and iterations are the most productive means at this stage. Focus should be placed on listing *all* tasks and their waste (e.g., waiting, rework, overprocessing), rather than on any task or flow optimization, which is performed in Future State mapping.

The Future-State Map: The Current-State Map displayed on the walls of a **war room** becomes the basis for iterative waste removal, including tailoring, and for improving task concurrency, synchronicity, precedence, and flow. It is critically important to plan for optimal precedence of both SE and PD tasks, showing which task is to feed which other task(s) with what data and when, understanding task dependencies and parent-child relationships. In order to shorten the program schedule, all tasks that can be worked on concurrently should be so planned, up to resource capacity. Experience indicates that some non-value-added tasks are self-evident and easy to remove, while others will require brainstorming and negotiations between the SE and other PD managers. Other wastes may yet never be discovered in a single program and become visible only in subsequent programs. Future-State VSM bears potential for huge direct payback, often measured in millions of dollars saved from a program per hour of mapping effort. Therefore, it should be performed as comprehensively as possible.

If the program follows a cadence of equal **takt periods**, the Future-State Map should next be parsed into the Periods (see Section 4.2).

The Current-State and Future-State value-stream maps may also be created concurrently, as follows. Map the process using the current understanding of the product development tasks, as each functional group understands its role. Then, walk the value stream to understand the need for the specifically identified tasks to ensure no waste enters the program. As the value stream is modified from the original current state, the future state is now developed—with better understanding of how each step contributes to the customer value.

The Work Breakdown Structure (WBS) and Gantt chart used for planning programs are typically of too low a resolution to serve as the final plan. Good planning should be done down to the level of individual self-contained tasks with clearly defined inputs and sources, outputs and destinations, and control and approval information.

SUGGESTED IMPLEMENTATION: Training in VSM for PD for Chief Engineers and other top managers.

LAGGING FACTORS: The culture of "we do not have time for this detailed planning, we must earn value as soon as possible."

SUGGESTED READING:

- J. Hoppmann, *The Lean Innovation Roadmap—A Systematic Approach to Introducing Lean in Product Development Processes and Establishing a Learning Organization*, Institute of Automotive Management and Industrial Production, Technical University of Braunschweig, Germany, 2009.

- H.L. McManus, *Product Development Value Stream Mapping Manual*, LAI Release Beta, Massachusetts Institute of Technology, LAI, April 2004.

- J.M. Morgan and J.K. Liker, *The Toyota Product Development System*. New York: Productivity Press, 2006.

- D. Lempia, *Using Lean Principles and MBE in Design and Development of Avionics Equipment at Rockwell Collins*, 26th *International Congress of the Aeronautical Sciences,* Anchorage, AK, September 16, 2008, paper 6.7.3.

- B. Oppenheim, Lean product development flow. *Journal of SE*, 2004, **7**, 4, pp. 352–376.

LEAN PRINCIPLE 2: Map the Value Stream (Plan the Program)

ENABLER: 2. Map the SE and PD Value Streams and Eliminate Non-Value-Added Elements.

SUBENABLER(S) AND USE RANKINGS:
  8. **Maximize concurrency of SE and other PD Tasks**. (U = 0.42)
  9. **Synchronize work flow activities using scheduling across functions, and even more detailed scheduling within functions**. (U = 0.65)
 10. **For every action, define who is responsible, approving, supporting, and informing (RASI), using a standard and effective tool, paying attention to precedence of tasks**. (U = 0.39)

VALUE PROMOTED: Shorter program schedule, robust Lean flow.

WASTE PREVENTED: Waiting and delays, rework, and other waste categories.

EXPLANATION: Systems Engineering (SE) is akin to a person's nervous system: sensing, controlling, and activating all program activities as needed in real time. Instead, in many traditional programs, SE is conducted remotely from other engineering and management activities. This results in inefficient sequential events, lack of buy-in to schedules and commitments throughout the PD team, unrealistic scheduling of tasks and long waits for task execution, receiving results late (often so late that output is obsolete upon arrival), long sequential approvals, and delayed release of results. The overall effect of these problems is excessive *cycle times*, poor coordination, and low quality of results. Value-added time of such an execution is a small fraction of overall *cycle time*, and rework is frequent because disjointed activities are poorly coordinated.

Subenabler 2.2.8 promotes maximum concurrency and synchronicity of SE and other PD tasks to minimize overall program time and the waste of waiting. In order to facilitate the maximum synchronicity of tasks, subenabler 2.2.9 promotes detailed scheduling of tasks both across functions (departments and suppliers) and within functions.

Subenabler 2.2.10 promotes the use of RASI-type tools for detailed task planning and allocation of resources. For each task, Systems Engineers in coordination with appropriate other PD engineers and managers should enter into an efficient and user-friendly RASI tool-specific names of the following individuals: the person responsible for the task execution, the person who will be approving the results, the individual(s) who may be needed to support the task work with knowledge or data, and the internal customer who will receive the output. Only one person should be listed in the "responsible" column by name, or no one will feel responsible. It is important that all these individuals coordinate task modalities, scope, and output before execution so that this task can be robustly executed *right the first time*. Users of RASI must pay attention to the proper precedence of tasks, so that a task output will not be scheduled as input in an earlier task.

In large programs operating with dynamically allocated resources it may be impossible to assign individual names early into RASI tool. In no cases should TBD be entered. A vastly better solution is to enter the name of the manager of an appropriate department, and that manager should serve as points of contact for task coordination until the time when a specific staff person is allocated.

SUGGESTED IMPLEMENTATION:
- Train team members in the benefits of planning for maximally concurrent and synchronous execution of tasks.
- Implement a user-friendly RASI tool (a number of tools are available commercially, or just use an Excel spreadsheet).
- Perform training in the RASI tool for all stakeholders.
- Mandate the use of the tool for all planning and scheduling of tasks.

LAGGING FACTORS:
- Inadequate, ad-hoc planning, traditional sequential and uncoordinated scheduling.
- Poor integration of SE with the rest of the program.
- Assumptions of "I thought you were doing that" throughout the program schedule and tasks.

SUGGESTED READING:
- J.M. Morgan and J.K. Liker, *The Toyota Product Development System*. New York: Productivity Press, 2006.
- D. Lempia, *Using Lean Principles and MBE in Design and Development of Avionics Equipment at Rockwell Collins*, 26th *International Congress of the Aeronautical Sciences*, Anchorage, AK, September 16, 2008, paper 6.7.3.

LEAN PRINCIPLE 2: Map the Value Stream (Plan the Program)
ENABLER: 2. Map the SE and PD Value Streams and Eliminate Non–Value-Added Elements.
SUBENABLER(S) AND USE RANKINGS: 11. Plan precisely to enable level workflow and schedule adherence, and to drive out arrival-time variation. (U = −0.30)
VALUE PROMOTED: Robust and predictable flow of work.
WASTE PREVENTED: Rework and waiting. Frustrations.
EXPLANATION: Accurate and dependable scheduling of tasks is crucial in PD to drive out arrival time variations. The accumulation of variation can cause serious schedule delay. Robust planning of a complex program requires that individual tasks be planned robustly and predictably, that is, with minimum uncontrolled *cycle time* and arrival-time variation. This predictability is achieved by standardization and leveling of tasks (smoothing of tasks effort in time). Smaller self-contained tasks are easier to plan than larger ones. Therefore, plans should break work into small predictable tasks that can be completed and integrated quickly, where practical. Subenabler 2.2.11 promotes precise planning and leveling of standardized tasks to assure predictable *cycle time* and arrival-time workflow at all times. Morgan and Liker [2006, p. 90] emphasize this aspect of planning: "At Toyota, schedule discipline means recognizing that intermediate dates are crucial to managing limited resources across multiple programs, and approaching these dates with rigor and precision. Toyota's suppliers are also aware of this tenacious attentiveness to intermediate target dates and would not consider letting a date slip. Toyota executes to schedule at a level of precision that is unknown in traditional product development systems. . . . Toyota engineers are often prepared to sleep on . . . mats at test equipment facilities if it will help complete test cycles on time."
SUGGESTED IMPLEMENTATION: An experienced and competent Chief Engineer (or equivalent role) should guide all detailed planning efforts according to the best industrial examples, in close coordination with competent functional and supplier managers, emphasizing consistent *cycle time* and arrival-time predictability.
LAGGING FACTORS: • Inadequate, ad-hoc planning. • Traditional sequential and uncoordinated scheduling. • Poor integration of SE with the rest of the program.
SUGGESTED READING: • J.M. Morgan and J.K. Liker, *The Toyota Product Development System*. New York: Productivity Press, 2006.

LEAN PRINCIPLE 2: Map the Value Stream (Plan the Program)
ENABLER: 2. Map the SE and PD Value Streams and Eliminate Non-Value Added Elements.
SUBENABLER(S) AND USE RANKINGS: 12. Plan below full capacity to enable flow of work without accumulation of variability and permit scheduling flexibility in work loading, i.e., have appropriate contingencies and schedule buffers. (U = −0.26)
VALUE PROMOTED: Robust, predictable, and consistent flow of work.
WASTE PREVENTED: Waiting, and schedule delay.
EXPLANATION: Just as freeway traffic slows down when the number of cars reaches road capacity, even in the absence of any accidents or road blocks, so workflow also slows when the amount of assigned work reaches organization capacity. This well-known fact from queueing theory can be explained by the accumulation of variability, as follows: It can be shown that variability does not average out to zero; instead the delays tend to accumulate. In order to eliminate slowdowns of workflow, good planners apply the following practices: • Scheduling work only up to about 90–95% of the theoretical capacity of the organization. • Demanding honest time estimates of task *cycle times* at the probability of meeting the estimate of 50%. • Including time buffers in the schedule. • Prioritizing tasks according to customer needs and corporate priorities. • Always maintaining a contingency plan (overtime, weekend buffers, or reserve resources). Note: The prioritization of tasks mentioned here is effective only if tasks are waiting for people, ready to be processed, with all inputs correct and at hand, and all task instructions clear. If these conditions are not satisfied (e.g., when people wait for correct inputs), prioritization becomes ineffective, and instead, other enablers focusing on making tasks ready for processing should be implemented first.
SUGGESTED IMPLEMENTATION: • Training of managers who plan workloads in their departments. • Training of SE who schedule and plan programs. • Training should be supplemented by reading of literature below. • Program scheduling with capacity reserves.

LAGGING FACTORS:

- Scheduling of work beyond the reasonable 90–95% of capacity.
- Lack of buffers and contingencies.
- The practice of padding *cycle time* estimates.

SUGGESTED READING:

- E.M. Goldratt, *Critical Chain*, The North River Press, 1997.
- J.M. Morgan and J.K. Liker, *The Toyota Product Development System*. New York: Productivity Press, 2006.

LEAN PRINCIPLE 2: Map the Value Stream (Plan the Program)

ENABLER: 2. Map the SE and PD Value Streams and Eliminate Non–Value-Added Elements.

SUBENABLER(S) AND USE RANKINGS:
13. **Plan to use visual methods wherever possible to communicate schedules, workloads, changes to customer requirements, and other information. (U = 0.22)**

VALUE PROMOTED: Higher quality of work, ability to react to defects in real time, visibility of delays, quality, and issues in real time.

WASTE PREVENTED: Rework, delays, inventory, overproduction, overprocessing.

EXPLANATION: Visual control boards monitoring the status and quality of work are used routinely on manufacturing floor of hi-tech companies. The same idea should be adopted in engineering environment. The best companies organize the work of departments, teams, and individuals for maximum direct visibility to all stakeholders, as follows:

Each engineer should use a small standardized dry-erase board hung in the most visible space just outside of his or her cubicle. The board should list all tasks at hand with deadlines and status, color coding each task (typically, green = progressing per plan; yellow = experiencing a manageable issue with no impact on the promised schedule; red = serious delay or issue, its brief description, and a summary of the corrective action undertaken). The marker board should be visible from outside a cubicle. The manager then can walk the corridors and see the real status of all work. So can other stakeholders (colleagues, customers, etc.). Managers should closely follow, mentor, and assist, if needed, with all yellow and red states. Managers should also monitor recurring red states and initiate corrective action to eliminate the issue once and for all, and enable *right the first time* executions. Such corrective actions may involve training, mentoring, and change of procedures, a *Kaizen* event, a Six Sigma project, or other actions. Note: The boards should not be implemented as computer screens because they will not be clearly visible at all times by all stakeholders. Managers must be trusted that they will not abuse the system to blame employees for imperfections (because then imperfections will be hidden, which obviates the purpose).

Departments and teams should have several marker boards displayed in a well-visible central location listing all current activities, using the same color coding as individuals. All important information should be displayed: e.g., the matrix of employees versus programs or teams they are serving; overall schedules of activities; trends in critical metrics, e.g., quality, *cycle time*, customer complaints; a list and status of issues identified and being corrected; status of employee certifications and training completed, and any other metric important to the success of the department.

Experience indicates that such visual controls are motivational to best work and conducive to immediate corrective action. Such controls promote teamwork and improve workflow speed and output quality.

These visual boards should not be implemented as computer screens: They lack the sufficient continuous visibility by all.

SUGGESTED IMPLEMENTATION:

- Explain the merits and examples of visual control boards to all employees.
- Assure employees that displaying all facts honestly in real time will not be used against them; in fact, they will be recognized for doing so. Eliminate fear and work to institute trust and teamwork.
- Then install boards and invite employees to share in the planning of information to be displayed. Mentor employees in their use.
- Demand frequent, regular, complete, and correct updating of information.
- React to new troubling information in a constructive way.
- Apply continuous improvement methods to eliminate each problem once and for all.

LAGGING FACTORS:

- Authoritarian management style, fear, culture of secrecy, lack of trust.

SUGGESTED READING:

- J.M. Morgan and J.K. Liker, *The Toyota Product Development System*. New York: Productivity Press, 2006.

LEAN PRINCIPLE 2: Map the Value Stream (Plan the Program)

ENABLER: 3. Plan for Front-Loading the Program.

SUBENABLER(S) AND USE RANKINGS:

1. **Plan to utilize cross-functional teams made up of the most experienced and compatible people at the start of the project to look at a broad range of solution sets. (U = 0.36)**
2. **Explore trade space and margins fully before focusing on a point design and too small margins. (U = 0.36)**

VALUE PROMOTED: Elegant, optimal, and least costly conceptual design that is conducive to trouble-free program execution and better value.

WASTE PREVENTED: Rework, waiting, overproduction, overprocessing, inventory.

EXPLANATION: The early conceptual phase of the program, often perceived as fuzzy or chaotic, has a major impact on the overall program success, cost and schedule. Usually, the worst approach is to rush a given design to a point solution prematurely. When such an unexplored point design is pursued under small design margins, its solution often appears impossible, leading to costly and wasteful iterations. It results in a non-optimum solution, requirements changes, or even system failure. Toyota design practice is not to allow a point solution until trade spaces and various constraints and limits have been fully explored first. Once the feasible region of a trade space is understood and defined, the optimum choice within that space is easy to find with zero to few iterations. Proper trade space exploration usually requires best experts from the domain and experts from all major subsystems to work together, looking at a broad range of solution sets, and brainstorming and negotiating together as a close, compatible team. Engineers often forget that iterations, particularly the ones that extend over several functions, are very costly and time consuming. They should be avoided whenever possible, and where indeed required, they should be optimized for the smallest and simplest possible loops.

Morgan and Liker [2006, p. 65] describe this approach at Toyota:
"Cross-functional teams made up of the most experienced people come together at the start of the project to look at a broad range of solution sets that anticipate and solve problems, to design countermeasures for quality and manufacturability, and to isolate inherent variability in product development in order to facilitate flawless execution in the next phase of the program."

SUGGESTED IMPLEMENTATION:
- Program plans must allocate enough time for front loading, in addition to time for planning.
- SE and enterprise management should organize domain experts and functional experts to work closely together during the conceptual phase of a program, with minimum bureaucracy.

LAGGING FACTORS:
- Management rushing engineers to a point solution without first exploring and understanding trade space.
- The customer insisting on a point solution prematurely.
- Domain experts not available during the conceptual design phase (are they busy creating next proposals?).

SUGGESTED READING:
- A.C. Ward, J.K. Liker, J.J. Cristiano, and D.K. Sobek II, The second Toyota paradox: How delaying decisions can make better cars faster. *Sloan Management Review*, Spring 1995a, **36**, 3, pp. 43–61.
- A.C. Ward, D.K. Sobek II, J.J. Cristiano, and J.K. Liker, Toyota, concurrent engineering, and set-based design, in J.K. Liker et al. (eds.), *Engineered in Japan: Japanese Technology Management Practices*. New York: Oxford Press, 1995b.
- J. Warmkessel, *Lean Engineering*. Lean Aerospace Initiative, MIT, http://lean.mit.edu, 2002.
- J.M. Morgan and J.K. Liker, *The Toyota Product Development System*. New York: Productivity Press, 2006.
- D. Lempia, *Using Lean Principles and MBE in Design and Development of Avionics Equipment at Rockwell Collins*, 26th *International Congress of the Aeronautical Sciences,* Anchorage, AK, September 16, 2008, paper 6.7.3.

LEAN PRINCIPLE 2: **Map the Value Stream (Plan the Program)**

ENABLER: **3. Plan for Front-Loading the Program.**

SUBENABLER(S) AND USE RANKINGS:
3. **Anticipate and plan to resolve as many downstream issues and risks as early as possible to prevent downstream problems. (U = 0.40)**

VALUE PROMOTED: Prevention, early resolution and mitigation of downstream issues and risks. Elimination of frustrations, delays, budget overruns, and even program failures.

WASTE PREVENTED: All waste categories, including program failure.

EXPLANATION: A wisely chosen Chief Engineer of the program (or equivalent role) should be experienced in the domain and in program management, "with scars from previous programs," who understands and anticipates major downstream issues and risks. The Chief Engineer should work proactively to mitigate all anticipated issues and risks to prevent any downstream problems.

SUGGESTED IMPLEMENTATION:
- Wisely select a Chief Engineer (or equivalent role) with expertise in the domain and enterprise.
- Create a corporate strategy for product development knowledge management.
- Build an excellent searchable database on intranet, with lessons learned in previous programs.
- Analyze applicable lessons learned before starting a new program.
- Disseminate lessons learned throughout the program, not just at post–mortem.
- As explained in Enabler 5.5, a Chief Engineer (or equivalent role) should be given RAA (responsibility, authority, and accountability) for program success. A Chief Engineer must have freedom to select teams, plans, and responsibilities for risk management and mitigation.

LAGGING FACTORS:
- Inexperienced Chief Engineer (or equivalent role).
- Management dissolved among several individuals, none of whom has the RAA.
- Poor transfer of lessons learned between programs.

SUGGESTED READING:
- B.R. Rich and L. Janos, *Skunk Works: A Personal Memoir of My Years at Lockheed*. Little, Brown and Company, Canada, 1994.
- A.C. Haggerty and E.M. Murman, *Evidence of Lean Engineering in Aircraft Programs, 25th International Congress of the Aeronautical Sciences*, Hamburg, Germany, September 2006.
- J.M. Morgan and J.K. Liker, *The Toyota Product Development System*. New York: Productivity Press, 2006.

LEAN PRINCIPLE 2: Map the Value Stream (Plan the Program)

ENABLER: 3. Plan for Front-Loading the Program.

SUBENABLER(S) AND USE RANKINGS:
4. Plan early for consistent robustness and *right the first time* under "normal" circumstances instead of hero behavior in later "crisis" situations. (U = 0.12)

VALUE PROMOTED: Smooth and predictable Lean flow of work through tasks.

WASTE PREVENTED: All categories of waste, including program failure, exceeded schedule and budgets.

EXPLANATION: It is a truism that the cheapest approach to any project is to do things *right the first time*; using predictable and robust processes and tools, and well-trained people. The so-called quality lever indicates that fixing a problem as early as possible in any program is vastly cheaper than doing it later. A notorious problem in recent programs is that schedule pressures cause engineers to rush through non-robust tasks and processes without optimizing them first, resulting in defects that, if left untended, lead to crises. Only when a crisis becomes visible, usually too late to mitigate it with only minor effort, programs tend to throw all resources at it, requiring costly and heroic behavior of engineers, managers, and experts. Such heroes are then recognized for saving the program. It is vastly better to use the heroes to build quality into processes and avoid crises in the first place.
In this consideration, it is important to distinguish between standard products and standard processes: In PD a product or work content is almost always unique. In contrast, most engineering processes should be standardized and optimized because they apply to many PD programs and should be predictable and robust, following best engineering practices. Note: The term *right the first time* means that if it is possible to achieve correct result on the first pass by proper planning, injection of experience and expertise, coordination, and communication, it should be done so to avoid rework waste. The term does not mean to imply that inherently iterative steps, such as trade space exploration, set-based approach, or architectural or design iterations should end after a single pass. But the *right the first time* philosophy should also apply even to those iterative tasks: preparing them well with knowledge, data, processes, and trained people, so that can be prepared and executed predictably.

SUGGESTED IMPLEMENTATION:

- An enterprise should allocate resources to capture the best engineering practice for each repeatable process, and then optimize the process and make it robust, predictable, consistent, and user friendly. This is a major challenge requiring Continuous Improvement (CI) tools and methods to be implemented effectively.
- Create a culture that rewards continuous improvement, not heroics.
- For improvement, prioritize any process that suffers from excessive uncontrollable variability.

LAGGING FACTORS:

- Budgeting of programs and enterprises that fail to allocate resources to make processes optimized and robust.
- Training of people not included in budgets.
- Rushing through tasks at the expense of quality.
- Lack of awareness that quality must be built into each process.

SUGGESTED READING:

- R. Leopold, *The Iridium Story: An Engineer's Eclectic Journey*, Minta Martin Lecture, MIT Department of Aeronautics and Astronautics, April 23, 2004.
- J.M. Morgan and J.K. Liker, *The Toyota Product Development System*. New York: Productivity Press, 2006.
- W.E. Deming, *Out of the Crisis*, Massachusetts Institute of Technology, Center for Advanced Engineering Study, 1982.

LEAN PRINCIPLE 2: Map the Value Stream (Plan the Program)

ENABLER: 4. Plan to Develop Only What Needs Developing.

SUBENABLER(S) AND USE RANKINGS:
1. **Promote reuse and sharing of program assets: Utilize platforms, standards, busses, and modules of knowledge, hardware, and software. (U = 0.32)**
5. **Maximize opportunities for future upgrades (e.g., reserve some volume, mass, electric power, computer power, and connector pins), even if the contract calls for only one item. (U = 0.40)**

VALUE PROMOTED: Development effort and cost spread among several programs, lower cost per program, faster schedule and faster response to customer needs, smaller and fewer risks, higher competitiveness of enterprises.

WASTE PREVENTED: All categories of waste are reduced by wise modularization and asset reuse.

EXPLANATION: The present subenablers (2.4.1 and 2.4.5) describe the wisdom of modularity. Companies that create a number of similar products/systems within a single domain (e.g., satellites, cars) should benefit from modularity and reuse of assets and lowering development costs. Engineering experience indicates that very few systems, if any, are so revolutionary that no prior assets, modules, or subsystems could be reused from earlier programs. The most typical reuse examples are designing new car models on a common platform or creating new satellite models on a common bus. Avionics boxes can be modularized, predesigned, and even prebuilt and pretested for use in several programs/models. Software can be created to be general enough for all the current and future functions and interfaces, with more channels than needed at the start. In fact, almost all subsystems in cars are modularized and reusable: engines, gearboxes, electronic subsystems, software, door handles, batteries, chassis, seats, radios, tires, etc. A car design program could not be competitive if all subsystems were designed from scratch. Yet, often, this is the practice in large governmental programs. Subsystems of similar satellites, busses, tanks, batteries, solar panels, antennas, computers, software, and so on, can and should be modularized.

Modularity applies to knowledge, hardware, and software. Modules intended for reuse should be created in a reasonably general way to serve not only the current program but also to anticipate future programs. In certain cases, it may be wise to reserve some volume, mass, electric power, computer power, and connector pins in a module, even if the contract calls for only one item, in order to maximize opportunities for future upgrades at minimum cost. There may be a small penalty in the volume, mass, and power of a module that has more capability than needed on a current program, but almost always the penalty will be more than compensated by the reduction in engineering labor and cost in the future systems. In the present economics of high technology, engineering labor is almost always the most expensive item in any program. In addition, availability of modules is often an advantage in marketing efforts for new programs.

SUGGESTED IMPLEMENTATION: Common assets benefit more than one program at once. Therefore, they should be coordinated at the corporate or enterprise level as a part of corporate strategy and marketing. Consider instituting an enterprise architect role that can look across organizations and past, present, and future programs and see opportunities for commonality and reuse. Ensure the architect role is staffed by a widely trusted and knowledgeable Systems Engineer with excellent communication and people skills. Encourage and share technology roadmaps and share with PD leaders. Good design of reusable modules calls for a degree of vision and understanding of future trends and market needs.

LAGGING FACTORS:
- Contracts (not uncommon in government programs) that require *the latest, greatest, and gold-plated*, disregarding existing or past assets, and demanding a wall of separation from other programs or corporate activities.
- Government contracts that do not allow expanding designs beyond the needs of the currently contracted system.

SUGGESTED READING:
- R. Leopold, *The Iridium Story: An Engineer's Eclectic Journey*, Minta Martin Lecture, MIT Department of Aeronautics and Astronautics, April 23, 2004.
- J.M. Morgan and J.K. Liker, *The Toyota Product Development System*. New York: Productivity Press, 2006.
- A.C. Haggerty and E.M. Murman, *Evidence of Lean Engineering in Aircraft Programs, 25th International Congress of the Aeronautical Sciences*, Hamburg, Germany, September 2006.
- E.M. Murman, *Lean Aerospace Engineering*, Littlewood Lecture AIAA-2008-4, January 2008.

LEAN PRINCIPLE 2: **Map the Value Stream (Plan the Program)**
ENABLER: **4. Plan to Develop Only What Needs Developing**.

SUBENABLER(S) AND USE RANKINGS:
2. **Insist that a module proposed for use is robust before using it**.
   (U = 0.20)
3. **Remove show-stopping research/unproven technology from critical**
   path, staff with experts, and include in the Risk Mitigation Plan.
   (U = 0.24)
4. **Defer unproven technology to future technology development efforts, or**
   future systems. **(U = 0.04)**

VALUE PROMOTED: Significantly shorter and more predictable program schedule and more robust designs.
WASTE PREVENTED: All categories of waste, including program failure.

EXPLANATION: Imagine a car assembly line in motion. After assembling
the chassis, a team looks to install an engine. But the engineers declare that
they have not designed the engine yet and need several years for it. The line
must stop and wait until the engine is designed. Absurd? But this is how
many governmental programs operate: waiting and consuming precious
program time until some high-risk, high-tech aspect of the program is
researched, designed, or resolved (even though the proposal promised no
low-TRL and only low-risk items). Let us jump to the ideal scenario,
practiced routinely in commercial programs. It has three phases: Research,
Development, and Design, as follows:
Research teams should address all items with low Technology Readiness Level
   (TRL) and bring them to the level of 5 or 6 using small co-located teams of
   experts working in dedicated laboratories. No large bureaucratic program is
   needed or should be used for this Research phase. Relevant experts should
   be engineers and scientists with PhDs, up to date with cutting-edge scientific
   literature, and a high degree of fascination with the challenge. The Research
   output should be in the form of knowledge that a technology for developing
   the subsystem/module is available, practical, and economical.
The *Development* phase then should take over from the Research team. Small
   teams of co-located experienced developmental engineers working within
   their relevant function(s) should then create Mature Robust Modules (MRM)
   of knowledge, or hardware, or software, including testing, reaching TRL 8
   or 9. The modules should be created with sufficient vision to serve a
   number of future programs.

The Design phase. Only when all modules are MRM (TRL level 8–9) should a system design proceed in accordance with fully developed Systems Engineering processes. If this correct approach is followed, design efforts can be reduced to tradeoffs of packaging MRM modules to satisfy all constraints of weight, space, power, longevity, etc., in an optimum way. This is the process used in car design, and it allows for complex new cars to be designed in under a year. If this sequence of Research, Development, and Design is not followed, programs will suffer from the costly absurd scenarios described above, risking schedule and cost penalties, requirements changes, and even program failure. Examples of such recent failed programs abound.

If a show-stopping research/unproven technology is detected in a program, it should be removed from critical path, staffed with experts, freed from the program bureaucracy, and included in the Risk Mitigation Plan. But a vastly better policy is to defer unproven technology to future technology development efforts, or future systems.

SUGGESTED IMPLEMENTATION:
- Maximum strategic enterprise emphasis and program-level emphasis on using the sequence Research, Development, and Design, as described here.
- An added emphasis in government acquisition system, demanding high TRL before issuing a design contract (post-milestone B contract), and harsh consequences if promises are broken.
- Using smaller, co-located and highly focused teams staffed with experts for the programs between acquisition milestones A and B.
- Mature, robust, and stable customer requirements.

LAGGING FACTORS:
- A large ambitious program starting with a low TR level (regardless of whether admitted in the proposal), which is not parsed properly into Research, Development, and Design phases, and/or mixing the three phases.

SUGGESTED READING:
- J. Young, *Memo to the Secretary of Defense*, Department of Defense, January 30, 2009.
- J.M. Morgan and J.K. Liker, *The Toyota Product Development System*. New York: Productivity Press, 2006.
- E.M. Murman, *Lean Aerospace Engineering*, Littlewood Lecture AIAA-2008-4, January 2008.

LEAN PRINCIPLE 2: Map the Value Stream (Plan the Program)
ENABLER: 5. Plan to Prevent Potential Conflicts with Suppliers.

SUBENABLER(S) AND USE RANKINGS:
1. **Select suppliers who are technically and culturally compatible.** (U = 0.46)
2. **Strive to develop seamless partnership between suppliers and the product development team.** (U = 0.21)
3. **Plan to include and manage the major suppliers as a part of your team.** (U = 0.42)
4. **Have the suppliers brief the design team on current and future capabilities during conceptual formation of the project.** (U = 0.13)

VALUE PROMOTED: Lean suppliers that provide a consistently perfect quality of supplied components, competitive and timely deliveries with just-in-time capabilities, fair price, world-class lead times and responsiveness, and partnering. Suppliers that are so good that no individual certifications or tests of parts are needed, and when required by contract, the tests always indicate "pass."

WASTE PREVENTED: All categories of waste, including program failure.

EXPLANATION: Modern high-technology industry has evolved into a horizontally integrated supply network, with estimated 60–95% or more of value provided by suppliers. Therefore, an excellent supply chain is absolutely critical to program success and company competitiveness. Until recently, the paradigm for supplier selection was simple: the lowest bid. During the last 10 years or so in the West (much earlier at Toyota and several other Japanese firms), an advanced body of knowledge called Supply Chain Management (or equivalent name) developed, demonstrating that vast benefits can be gained by all stakeholders when critical suppliers are in a strategic, long-term seamless partnership with a prime contractor. Their mutual relationships are based on trust, honesty, openness, teamwork, good communications, and mutual interest. Hostility, confrontation, and exploitation should not exist in modern supplier networks. "Supply chains must be planned (designed) or one may not like the outcome of no planning" [Deloitte, 2003]. The supply chain management method distinguishes between strategic suppliers, critical suppliers, important suppliers, and commodity (e.g., toilet paper) suppliers (other similar terms may be used for the characterizations). The four above subenablers (2.5.1–2.5.4) address the ideal supply network. Strategic, critical, and important suppliers must be selected for long-term partnership, based on past performance, competence and expertise, competitiveness, training of employees, cultural compatibility with the prime contractor (e.g., both having implemented Lean), willingness to cooperate closely at many levels, participation in design reviews, briefing design teams on the latest technologies and capabilities, and generally acting as if a supplier's team were a seamless part of a buyer's design team. The goal to the buyer is to achieve consistently perfect quality of supplied components, competitive and timely deliveries with just-in-time capabilities, world-class lead times, responsiveness, and excellent service. The goal for the supplier is to have a steady long-term buyer who pays a fair price and makes timely payments. All suppliers should also receive crystal-clear specifications and requirements for their products and precise descriptions of all tests their products will have to pass.

The use rankings of these subenablers indicate a less than perfect level of. implementation in industry. These subenablers offer a significant potential for productivity increase and waste elimination.

A few quotes from the literature on suppliers illustrate these points:

- "Only 7% of companies today are effectively managing their supply chain. However, these companies are 73% more profitable than other manufacturers." [Deloitte, 2003]

- "Xerox and other leading companies have reconceived traditional customer-supplier relationships. Price negotiation is no longer the only or most important dimension of these relationships.... [Xerox] regards suppliers as partners or joint venture associates in its business; consequently, Xerox has cut its supplier base from more than 5000 in the early 1980s to 420 in 1993." [Bogan, 1994, p. 175]

- Toyota upholds the idea that "it is more expensive in the long run to pick the cheapest supplier if this supplier is not ready to meet your requirements." [Morgan and Liker, 2006, p. 192]

- [The underlying purpose of developing seamless partnership with suppliers and the product development team is that the customer cannot tell the difference between the two]. "Toyota recognizes this and makes sure that every car part reflects Toyota quality. To achieve this, they [sic] make every supplier an extension of Toyota's PD process and lean logistics chain. Toyota delegates tasks to suppliers, but ultimately, it is Toyota that is fully responsible and fully accountable for all subsystems as well as the final product. Outsourcing does not absolve Toyota of responsibility and accountability." [Morgan and Liker, 2006, p. 180]

- "When Toyota invites a supplier to send guest engineers, it is a significant commitment to long-term co-prosperity. The supplier knows they have [sic] earned a long-term place in the Toyota enterprise. These are highly coveted positions. The supplier will become intimately familiar with Toyota's product development practices and get advanced information on new model programs." [Morgan and Liker, 2006, pp. 193–194].

SUGGESTED IMPLEMENTATION:
- Have the top Purchasing Manager of the buying company (at corporate level) acquire the state-of-the-art education and knowledge in modern Lean Supply Chain Management. Normally, this would be a graduate degree in Supply Chain Management (or similar name).
- Have the manager implement the strategic management of the supply chain, if necessary with assistance of experts.
- Train other stakeholders (engineers and managers who interact with suppliers) in the above mentioned characteristics to build a common culture of cooperation, teamwork, and partnership.
- Patiently develop relations with suppliers that are based on openness, teamwork, good communications, mutual interest, fairness, and long-term partnership. But demand Lean from all suppliers.

LAGGING FACTORS:
- Selecting suppliers by lowest bid.
- Too many suppliers for like items.
- Lack of corporate awareness of education and knowledge of modern supply chain management.
- Lack of trust between suppliers and buyers.
- Specifications given to a supplier unfinished, incomplete, incorrect.

SUGGESTED READING:
- C. Bogan, *Benchmarking Best Practices*, McGraw Hill, 1994.
- J.M. Morgan and J.K. Liker, *The Toyota Product Development System*. New York: Productivity Press, 2006.
- A.C. Haggerty and E.M. Murman, *Evidence of Lean Engineering in Aircraft Programs, 25th International Congress of the Aeronautical Sciences*, Hamburg, Germany, September 2006.
- R. Leopold, *The Iridium Story: An Engineer's Eclectic Journey*, Minta Martin Lecture, MIT Department of Aeronautics and Astronautics, April 23, 2004.
- E.M. Murman, T. Allen, K. Bozdogan, J. Cutcher-Gershenfeld, H. McManus, D. Nightingale, E. Rebentisch, T. Shields, F. Stahl, M. Walton, J. Warmkessel, S. Weiss, and S. Widnall, *Lean Enterprise Value: Insights from MIT's Lean Aerospace Initiative*, Palgrave, Hampshire, 2002.

LEAN PRINCIPLE 2: Map the Value Stream (Plan the Program)

ENABLER: 6. Plan Leading Indicators and Metrics to Manage the Program.

SUBENABLER(S) AND USE RANKINGS:
1. **Use leading indicators to enable action before waste occurs**. ($U = -0.04$)
2. **Focus metrics around customer value, not profits**. ($U = -0.33$)
3. **Use only few simple and easy to understand metrics and share them frequently throughout the enterprise**. ($U = 0.16$)
4. **Use metrics structured to motivate the right behavior**. ($U = 0.00$)
5. **Use only those metrics that meet a stated need or objective**. ($U = 0.04$)

VALUE PROMOTED: Measuring program performance realistically, robustly, and having the ability to undertake timely changes when needed.

WASTE PREVENTED: All categories of waste, including program failure.

EXPLANATION: Recent programs are notorious for massive bureaucracy, serving program management and customers with complex controls and metrics. A vicious feedback loop evolves: Poorly progressing programs stimulate more controls on the part of management and government, which insist on more metrics and bureaucracy, which consume more resources, thus further slowing programs, promoting even more controls, and so on. As discussed in Chapter 5, on average, 12% of program charged time is spent on value-creating activities and 88% is wasted. Many metrics presently used in industry may indicate good progress, while every employee knows that the program has actually stalled or is failing. This condition is unsustainable. The solution is not more bureaucracy or controls. The present subenablers (2.6.1–2.6.5) promote going back to common sense. They promote the use of leading indicators to detect unhealthy trends before actual problems occur, when such problems are actually easy to correct. The subenablers promote using a few simple-to-understand metrics that measure what should be measured: customer value (rather than profits), conditioning the right behavior, and serving a legitimate need of the program control and management rather than the self-serving needs of some managers. Metrics should be shared with the entire program community. When designing metrics, it is important to keep in mind that customers need four performance qualifiers: quality, promised price, dependable schedule, and customer-friendly service. All other aspects are less important or irrelevant, or derivative from those. Note the use rankings for these subenablers are poor, indicating large room for improvement.

SUGGESTED IMPLEMENTATION:
- Top program and corporate managers should acquire the knowledge of good metrics.
- The government should insist on low-bureaucracy and simple and effective metrics
- Every program should follow low-bureaucracy and simple and effective metrics.
- The measured parameters should be openly disseminated to teams.

LAGGING FACTORS:
- Bureaucratic, useless metrics.
- *Cost-plus* contracting.
- Dissolved management of programs.

SUGGESTED READING:
- D.H. Rhodes, R. Valerdi, and G.J. Roedler, *Systems Engineering Leading Indicators for Assessing Program and Technical Effectiveness*. Wiley Periodicals, Inc., 2008.
- J.E. Cunningham, O.J. Fiume, and E. Adams, *Real Numbers: Management Accounting in a Lean Organization*. Managing Times Press, 2003.
- *Agreement between New United Motor Manufacturing, Inc. (Toyota NUMMI) and the United Auto Workers* (union), August 1, 1998.
- J.M. Morgan and J.K. Liker, *The Toyota Product Development System*. New York: Productivity Press, 2006.

7.2.3   Lean Principle 3: Flow

Summary   In complex programs, opportunities for program progress to stop are overwhelming, and it takes careful preparation, planning, and coordination effort to keep flows going. The third Principle, "Flow," includes those practices that enable work to flow smoothly and continuously, minimizing stopping, waiting, rework, or backflow.

Eight enablers are listed under this Principle, as follows. The numbers in parentheses list the number of subenablers under each enabler.

Enabler 3.1 Execute the Program According to the INCOSE Handbook Process.

The first enabler listed under each Principle emphasizes the same fundamental point: to add the wisdom of Lean Thinking to the traditional Systems Engineering process rather than replace the process. So, again, the present enabler recommends continuing to use the INCOSE SE Handbook, or any other equivalent SE handbook, and in addition to implement the enablers that follow.

Enabler 3.2 Clarify, Derive, and Prioritize Requirements Early and Often During Execution. (6)

In contrast to highly successful early aerospace programs that were defined with only a few customer-level requirements, the present trend is to specify many hundreds or even thousands of customer-level requirements. This large number of requirements necessitates careful and continuous effort to capture, clarify, derive, and prioritize them during program execution, as well as dedicated communication channels for requirement clarifications. If this is not done well, the subsequent flow of the program will suffer from delays, rework, and other types of waste, with negative consequences on budget and schedule. The six subenablers are helpful in preventing these wastes.

Enabler 3.3 Front-load Architectural Design and Implementation. (5)

An early exploration of multiple concepts, architectures, and designs is a critical aspect of lean PD. This is conducive to optimum solutions that remain stable throughout the program and shorten the path to a final design. The five subenablers promote and explain the various frontloading practices.

Enabler 3.4 Systems Engineers to Accept Responsibility for Coordination of PD Activities. (4)

The four subenablers promote excellent coordination of PD activities, leading to excellent teaming between SE engineers and other PD engineers.

Enabler 3.5 Use Efficient and Effective Communication and Coordination. (9)

These nine subenablers promote efficient and effective communications among stakeholders in PD programs. This is a critical aspect of well functioning PD. The amount of communications must be just right: Too much bad communications (e.g., unneeded emails) takes away from value-creating time, and too little causes misunderstandings, rework, and other wastes.

Enabler 3.6 Promote Smooth SE Flow. (6)

The six subenablers provide a checklist of good practices that enable smooth flow of work. They include frequent *integrative events* for resolutions of issues, challenging customers on technical grounds for program stability, minimizing handoffs, optimizing human resources for best creation of value, using consistent measurements standards, and mutual understanding of needs.

Enabler 3.7 Make Program Progress Visible to All. (4)

The four subenablers promote the adoption of a successful practice of displaying all critical information in public spaces visible to all. Such information is motivational to immediate corrective action. The information displayed should include work status, trends of important parameters, status of employee training, traffic light system, and making all imperfections visible.

Enabler 3.8 Use Lean Tools, (5)

The five subenablers promote the use of lean tools that maximize work flow, reduce non–value-adding efforts, and help engineers in the creation of value.

The following tables describe the Principle 3 subenablers individually.

LEAN PRINCIPLE 3: Flow

ENABLER: 2. Clarify, Derive, Prioritize Requirements Early and Often During Execution.

SUBENABLER(S) AND USE RANKINGS:

1. **Since formal written requirements are rarely enough, allow for follow-up verbal clarification of context and need, without allowing** *requirements creep*. (U = 0.36)
2. **Create effective channels for clarification of requirements (possibly involve customer participation in development IPTs).** (U = 0.56)
3. **Listen for and capture customer** *unspoken requirements*. (U = 0.20)
4. **Use architectural methods and modeling for system representations (3D integrated CAE toolset, mockups, prototypes, models, simulations, and software design tools) that allow interactions with customers as the best means of drawing out customer requirements.** (U = 0.72)

VALUE PROMOTED: All aspects of value.

WASTE PREVENTED: All waste categories, including program failure.

EXPLANATION: It is self-evident that good requirements are a necessary condition for program success. Yet, it is not widely appreciated that written requirements alone are not sufficient to capture the value proposition completely or correctly for three reasons. The first reason is linguistic: A requirement is a sentence with several words in a natural language, each word having multiple meanings. The number of combinations of meanings of this many words in a sentence is large. Thus, a requirement written in a natural language is subject to inherent linguistic ammbiguities. The second reason is the psychological handoff: The person writing the requirements has in his or her mind the rich context of the system needs and operations. Out of this context he or she formulates the few words of the requirement into a new document with a new context. The person reading the requirement sees only those few words in a different context and has no access to the original mindset of the writer. Experienced Systems Engineers understand that the risk of information loss in such handoffs is significant. The third reason is completeness: Experience indicates that written requirements are rarely capable of capturing information in its entirety, including need, context, understanding of the user's community and culture, operational scenarios, life cycle challenges, and many others. The history of technology is full of failures because some aspect of need was ignored in requirements development. In summary, written requirements alone must always be assumed as imperfect, and therefore requiring follow-up clarifications. Subenabler 3.2.1 promotes verbal follow-up to written requirements, emphasizing that it should be performed without *requirements creep*, which in practice means by properly trained, competent, and authorized staff. Subenabler 3.2.2 promotes the establishment of effective organizational communications channels for clarifications, such as designated pairs of experienced employees for each requirement (the writer and the expert on the receiving side).

In spite of best efforts, in many complex programs, customer stakeholders do not have sufficient expertise or experience to formulate all requirements explicitly, correctly, and completely. Subenabler 3.2.3 places the burden of capturing the *unspoken requirements* on contractors. Each contractor should treat the capturing of *unspoken requirements* as an ethical and professional obligation. "*Unspoken requirements* are like expected requirements in the sense that customers do not feel obligated to tell the developer about them, either because they feel they are clear and obvious or because they do not know that they exist." [Kamrani and Salhieh, 2002, p. 91].

Subenabler 3.2.4 promotes the best means for drawing requirements during interactive sessions between contractors and users/customers: prototypes, models, CAE tools, simulations, and so on. Such interactive tools are conducive to better understanding and capturing of need and value and for closer cooperation between customer and contractor representatives.

SUGGESTED IMPLEMENTATION:

- Train all relevant stakeholders to understand that even the best written requirements are not enough and that additional clarification effort will be needed and must be planned for.

- Train employees in requirements follow-up practices that are legal, effective, and do not cause *requirements creep*.

- Establish effective channels for the follow-up. The best channels are pairs of designated employees: the requirement writer and a competent reader.

- Provide resources (time, staff, and process) for capturing the *unspoken requirements* and for follow-up.

- Excellent human relations based on trust, openness, honesty, commonality of goals, and competence are needed for success.

- The above cannot possibly do justice to the huge body of knowledge that is needed to properly capture requirements. All relevant stakeholders must be well trained in the job: user communities, customer stakeholders, contractor's stakeholders, systems engineers, design and test engineers, and managers at all levels. Corporate support is needed also. Every effort should be devoted to this extensive task because if not done right, the program will surely suffer, and may even fail. A number of large, recent programs have been cancelled because they wasted billions of dollars partly due to ill formulation of requirements.

- Increasingly, tools are available to measure and aid in requirements formulation and changes, such as the Requirements Stability Index (RSI) to control *requirements creep*.

LAGGING FACTORS:

- Traditional attitude "the customer is responsible for the requirements and we only execute as directed."
- *Cost-plus* contracting that de-incentivizes the effective capture of value proposition early.

SUGGESTED READING:

- J.M. Morgan and J.K. Liker, *The Toyota Product Development System*. New York: Productivity Press, 2006.
- E.M. Murman, *Lean Aerospace Engineering*, Littlewood Lecture AIAA-2008-4, January 2008.
- R. Leopold, *The Iridium Story: An Engineer's Eclectic Journey*, Minta Martin Lecture, MIT Department of Aeronautics and Astronautics, April 23, 2004. C.M. Jones, 2006.
- A.K. Kamrani and S.M. Salhieh, *Product Design for Modularity*, Norwell, MA: Kluwer Academic Publishers, 2002.
- A.C. Haggerty and E.M. Murman, *Evidence of Lean Engineering in Aircraft Programs, 25th International Congress of the Aeronautical Sciences*, Hamburg, Germany, September 2006.

LEAN PRINCIPLE 3: Flow

ENABLER: 2. Clarify, Derive, Prioritize Requirements Early and Often During Execution.

SUBENABLER(S) AND USE RANKINGS:
5. *Fail early-fail often* **through rapid learning techniques (prototyping, tests, digital preassembly, spiral development, models, and simulation). (U = 0.04)**

VALUE PROMOTED: Reduce overall program cost and time by exploring trade space using efficient trial-and-error iterations early in program execution. Fast iterate and learn as much as possible, as early as possible, while it is still affordable.

WASTE PREVENTED: Defects, waiting, overprocessing, overproduction, transportation.

EXPLANATION: "Empirical evidence shows that poor decisions early in the process have a negative impact on cost and timing, which increase exponentially as time passes and project matures. Although this is generally recognized, very few companies understand how to take advantage of the golden front-end opportunity by making wise front-end investments. Toyota is one of the few." [Morgan and Liker, 2006, p. 39].
Implementing quick and inexpensive trial-and-error iterations in early program phases permits exploration of a large number of alternatives while costs are orders of magnitude lower than they would be in later stages. This approach converges on an optimum solution earlier and makes it cheaper. Furthermore, it is conducive to a steady and robust work flow during subsequent program flows. Iterations should be planned and budgeted accordingly. Needed functions should perform the planning together for optimum cross-function iterations.
A number of modern technologies are available to execute iterations effectively: rapid prototyping, rapid testing, digital preassembly, modularity, models, simulations, and spiral development.
As Morgan and Liker [2006, p. 82] point out: "By aligning objectives across functions and developing designed-in countermeasures, Toyota enables synchronized, cross-functional process flow and eliminates one of the mortal enemies of flow in product development—*unscheduled and late engineering changes*. These engineering changes disrupt the process, drive excessive cost, and negatively impact quality."
Note: Subenabler 3.2.5 is closely linked to subenabler 3.3.1: "Explore multiple concepts, architectures and designs early."

SUGGESTED IMPLEMENTATION: Training of engineers and managers in the benefits and efficient usage of early trade space exploration.

LAGGING FACTORS:
- Jumping to a point-design too early without exploring the trade space.
- *Earned-value* pressures to show tangible progress early.
- False sense of making quick initial progress without exploring alternatives.

SUGGESTED READING:
- J.M. Morgan and J.K. Liker, *The Toyota Product Development System*. New York: Productivity Press, 2006.
- E.M. Murman, *Lean Aerospace Engineering*, Littlewood Lecture AIAA-2008-4, January 2008.
- D. Lempia, *Using Lean Principles and MBE in Design and Development of Avionics Equipment at Rockwell Collins, 26th International Congress of the Aeronautical Sciences,* Anchorage, AK, September 16, 2008, paper 6.7.3.
- A.C. Haggerty and E.M. Murman, *Evidence of Lean Engineering in Aircraft Programs, 25th International Congress of the Aeronautical Sciences*, Hamburg, Germany, September 2006.
- R. Leopold, *The Iridium Story: An Engineer's Eclectic Journey*, Minta Martin Lecture, MIT Department of Aeronautics and Astronautics, April 23, 2004.

LEAN PRINCIPLE 3: Flow

ENABLER: 2. Clarify, Derive, Prioritize Requirements Early and Often During Execution.

SUBENABLER(S) AND USE RANKINGS:
6. Identify a small number of goals and objectives that articulate what the program is set up to do, how it will do it, and what the success criteria will be to align stakeholders—and repeat these goals and objectives consistently and often. (U = 0.28)

VALUE PROMOTED: Constant focus on and alignment with the main goals and objectives of value propositions.

WASTE PREVENTED: All types of waste, primarily rework and waiting.

EXPLANATION: The present corporate culture tends to compartmentalize engineers through office architecture (people working in their individual cubicles at their computers), geographical distribution, and narrow technical tasks. In addition, many engineers are employed on only relatively short tasks in much longer programs. This environment is not conducive to good understanding of a *big picture*: what a program is all about, what the value is of the system being created, and which customer needs are being served. Without this knowledge, engineers are unlikely to produce good value. At best they will satisfy minimum requirements. Good managers understand this. They identify a small number of goals and objectives that articulate what a program is set up to do, how it will do it, and what the success criteria will be to align stakeholders—and repeat these goals and objectives consistently and often to all stakeholders. This is particularly important in large, long, distributed programs.

SUGGESTED IMPLEMENTATION: This is one of the easiest subenablers to implement: Just follow the subenabler instruction and insist on repeating these goals and objectives consistently and often. Instruct new hires immediately about the program goals and objectives.

LAGGING FACTORS:
- Geographical distribution of program stakeholders.
- Compartmentalized office architecture and culture.
- Overutilization of people leaving no time to think of what is really important; "can't see the forest for the trees."

SUGGESTED READING:
- R. Leopold, *The Iridium Story: An Engineer's Eclectic Journey*, Minta Martin Lecture, MIT Department of Aeronautics and Astronautics, Apr. 23, 2004.
- J.M. Morgan and J.K. Liker, *The Toyota Product Development System*. New York: Productivity Press, 2006.
- A.C. Haggerty and E.M. Murman, *Evidence of Lean Engineering in Aircraft Programs, 25th International Congress of the Aeronautical Sciences*, Hamburg, Germany, September 2006.
- E.M. Murman, *Lean Aerospace Engineering*, Littlewood Lecture AIAA-2008-4, January 2008.

LEAN PRINCIPLE 3: Flow

ENABLER: 3. Front Load Architectural Design and Implementation.

SUBENABLER(S) AND USE RANKINGS:
1. **Explore multiple concepts, architectures and designs early.** (U = 0.44)
2. **Explore constraints and perform real trades before converging on a point design.** (U = 0.46)

VALUE PROMOTED: Frontloading shortens overall program time, while lowering total cost and improving quality.

WASTE PREVENTED: All types of waste.

EXPLANATION: An early exploration of multiple concepts, architectures, designs, and constraints is a critical aspect of frontloading. This is conducive to finding optimum solutions that remain stable and robust throughout a given program. Frontloading seems counterintuitive because it requires a larger effort to be performed early, appearing to slow the progress without producing easy results, while an immediate solution is already in sight. Many budget-focused managers rebel against spending program resources early and instead push to a point design. Also, many contracts governed by the *earned-value* bookkeeping promote quick focus on deliverables. They are wrong. In fact, it is the opposite approach of jumping to a preconceived notion or a point design too early, without understanding the entire tradespace and constraints—which often leads to massive cross-functional iterations and design changes in later design stages when iterations are dramatically more costly and time consuming. Experienced engineers understand the benefits of frontloading.

Tradespace exploration of an entire system also prevents suboptimum solutions of incompatible subsystems and parts.

A particular approach to frontloading involves exploration of alternatives to systematically narrow the trade space down to a single, optimum final choice. This approach is known as "set-based concurrent engineering," a term coined by an academic group of authors from the University of Michigan (A. C. Ward, J. K. Liker, J. J. Cristiano, and D. K. Sobek II). Set-based concurrent engineering is usually contrasted to iterative point-based design. Iterating design when starting from a point design carries the risk that the number of iteration loops is unknown and could consume a significant portion of budget and schedule. And, in extreme cases, this may require changes to requirements. In contrast, set-based approaches predetermine the number of sets to evaluate. See also subenabler 3.2.5.

SUGGESTED IMPLEMENTATION:

- Training of engineers and managers in the benefits and methods of efficient early trade space exploration.
- Expert managers should lead the trade space exploration effort.

Morgan and Liker [2006, pp. 48–51] propose the following approach to the set-based design (paraphrased):

- Intentionally identify multiple solutions to design problems before selecting just one.
- Encourage engineers (both upstream and downstream) to discuss alternatives early before a fixed decision has been reached on a single design from one perspective.
- Use set-based tools such as tradeoff curves to identify the trade-offs of various solutions from different perspectives.
- Capture past knowledge in checklists in the form of graphs and equations that show the effects of different alternatives.
- Use system methods like parametric design that quickly show system impacts when parameters are changed.

An example of Toyota's lean PD process is the Prius hybrid. "At the time this model was being developed, there was intense pressure to meet aggressive time lines set by the company president, Mr. Okuda. The timeline was shortened and shortened despite the fact that the vehicle was to have an entirely new power train. ... The simplest solution for the chief engineer, Mr. Uchiyamada, was to short-circuit the process. But Mr. Uchiyamada was a true Toyota man, a chief engineer whose father was also a chief engineer. He refused to compromise, insisting on adherence to the established Toyota process of considering broad alternatives and gradually narrowing the alternatives until a superior engine, body style, and transmission could be selected" [Morgan and Liker, 2006, p. 48].

LAGGING FACTORS:

- Jumping to a point-design too early without exploring trade space first.
- False sense of making quick progress without exploring alternatives.
- *Earned-value* pressures to show tangible progress early.

SUGGESTED READING:
- J.M. Morgan and J.K. Liker, *The Toyota Product Development System*. New York: Productivity Press, 2006.
- J.K. Liker, D.K. Sobek II, A.C. Ward, and J.J. Cristiano, Involving suppliers in product development in the US and Japan: Evidence for set-based concurrent engineering. *IEEE Transactions in Engineering Management*, May 1996, **43**, 2, pp. 214–240.
- D.K. Sobek II, A.C. Ward, and J.K. Liker, Toyota's principles of set-based concurrent engineering. *Sloan Management Review*, Winter 1999, **40**, 2, pp. 67–83.
- A.C. Ward, J.K. Liker, J.J. Cristiano, and D.K. Sobek II, The second Toyota paradox: How delaying decisions can make better cars faster. *Sloan Management Review*, Spring 1995a, **36**, 3, pp. 43–61.
- C. Ward, D.K. Sobek II, J.J. Cristiano, and J.K. Liker, Toyota, concurrent engineering, and set-based design, a chapter of Liker et al. (eds.), *Engineered in Japan: Japanese Technology Management Practices*. New York: Oxford Press, 1995b, pp. 192–216.
- D. Lempia, *Using Lean Principles and MBE in Design and Development of Avionics Equipment at Rockwell Collins*, 26th *International Congress of the Aeronautical Sciences,* Anchorage, AK, September 16, 2008, paper 6.7.3.

LEAN PRINCIPLE 3: Flow

ENABLER: 3. Front Load Architectural Design and Implementation.

SUBENABLER(S) AND USE RANKINGS:
3. Use a clear architectural description of the agreed solution to plan a coherent program, engineering and commercial structures. (U = 0.44)

VALUE PROMOTED: Clear architectural description of the agreed solution is conducive to subsequent robust and efficient flow and trouble-free program. This, in turn, tends to reduce waste, cost and schedule, and promotes satisfaction of stakeholders.

WASTE PREVENTED: All waste categories.

EXPLANATION: Lack of coherent design, or lack of consensus on baseline design can torpedo the entire program. Best programs achieve a clear architectural description of the final agreed solution as a major milestone of frontloading. Such architectures tend to capture critical elements and interfaces of the system and present them in a relatively easy-to-follow manner, with high level of coherency. Such architecture is easy to share among program teams and is conducive to error-free interpretations and detailed planning. Good architecture is also adaptable to efficient changes.

SUGGESTED IMPLEMENTATION: Promote the use of Systems Architecting in the conceptual phase of the program. Use System Architects experienced in program domain. Demand that a clear architectural description representing consensus of major stakeholders be completed during an early phase of the program and not as an afterthought. Then disseminate the architecture to program team for subsequent detailed program planning and execution. Share applicable elements of the architecture with critical suppliers.

LAGGING FACTORS:
- Lack of culture of architecting the design.
- Lack of systems architects experienced in the domain.
- Schedule pressures driven by *earned value*.

SUGGESTED READING:
- J.M. Morgan and J.K. Liker, *The Toyota Product Development System*. New York: Productivity Press, 2006.
- D. Lempia, *Using Lean Principles and MBE in Design and Development of Avionics Equipment at Rockwell Collins*, *26th International Congress of the Aeronautical Sciences*, Anchorage, AK, September 16, 2008, paper 6.7.3.

LEAN PRINCIPLE 3: Flow

ENABLER: 3. Front Load Architectural Design and Implementation.

SUBENABLER(S) AND USE RANKINGS:
4. All other things being equal, select the simplest solution. (U = 0.12)

VALUE PROMOTED: Reduced unneeded complexity, cost, time, and
frustrations.

WASTE PREVENTED: All waste types.

EXPLANATION: Three well-known mantras apply here:
1. "Any fool can make anything complex but it takes a genius and courage to
   create a simple solution."—Albert Einstein
2. "KISS = keep it simple, stupid"—a timeless engineering mantra.
3. "Occam's Razor = one should not increase, beyond what is necessary, the
   number of entities required to explain anything."

Driven by competitive pressures, corporate hunger for large programs, and
frequent government demand for *the latest, greatest, and gold-plated*, large
companies often pursue excessive options, features, and performance
envelopes. Often, the majority of these features will never be used. The
excessive number of features is also notorious in commercial software. The
implementation of too many features results in excessively complex
systems, complex software and logistics, long and costly testing, difficult
revisions, and complex controls. Such systems can consume budget and
schedule before the program is completed. These programs then face the
choice to either cut corners on testing or extend program schedule and
budget. Neither solution has objective merits. Altogether, there is great merit
in "keeping things simple, all else being equal."

SUGGESTED IMPLEMENTATION: Chief Engineers (or equivalent role)
should emphasize the importance of seeking the simplest solutions, all other
things being equal.

LAGGING FACTORS:
- Seeking excessive complexity as an element of value.
- The hunger for large programs, and the government's demand for *the latest,
  greatest, and gold-plated*.

SUGGESTED READING:
- J.K. Liker, *The Toyota Way: 14 Management Principles from the World's
  Greatest Manufacturer*. McGraw-Hill Professional, 2004.
- J.M. Morgan and J.K. Liker, *The Toyota Product Development System*. New
  York: Productivity Press, 2006.
- D. Lempia, *Using Lean Principles and MBE in Design and Development of
  Avionics Equipment at Rockwell Collins*, 26th *International Congress of the
  Aeronautical Sciences,* Anchorage, AK, September 16, 2008, paper 6.7.3.

LEAN PRINCIPLE 3: Flow
ENABLER: **3. Front Load Architectural Design and Implementation**.
SUBENABLER(S) AND USE RANKINGS: **5. Invite suppliers to make a serious contribution to SE, design, and development as program trusted partner**. (U = 0.24)
VALUE PROMOTED: Better and more competitive systems created, less costly, fewer frustrations in relations with suppliers, better common understanding of system, fewer problems in flow.
WASTE PREVENTED: Waiting, overprocessing, overproduction, transportation.
EXPLANATION: As stated earlier, in modern systems, 60–95% or more of system value is provided by suppliers. Thus, a healthy, mutually beneficial partnership between buyers and suppliers, growing the business together, is critical to long-term success. In such a partnership, suppliers have a vested interest in providing the best possible solutions to the buyer, so that together they create a competitive system and grow the market. The suppliers usually have a deeper and more recent knowledge of ordered subsystems than the buyer. Trusted suppliers should share this knowledge with the buyer, as a natural contribution to the partnership. Specifically, the buyer should invite all strategic and critical suppliers to make a serious contribution to SE, design, and development, helping to evaluate possible options, technical and cost tradeoffs, and risks. An ideal interaction should appear seamless: Supplier engineers should work with buyer's engineers as a co-located team designing the system together. Toyota was one of the earliest corporations developing powerful lean supply network: "When Toyota started building automobiles, it did not have capital or equipment for building the myriad of components that go into a car. One of Eiji Toyoda's first assignments as a new engineer was to identify high-quality parts suppliers that Toyota could partner with. At that time they did not have the volume to give a lot of business to suppliers. In fact, some days they did not build a single vehicle because they did not have enough quality parts. So Toyoda understood the need to find solid partners. All that Toyoda could offer was the opportunity for all partners to grow the business together and mutually benefit in the long term. So, like the associates who work inside Toyota, suppliers became part of the extended family who grew and learned the Toyota Production System" [Liker, 2004, p. 202].

Being a partner is the highest level of Toyota's tier structure for its suppliers. Toyota partners such as Denso, Araco, and Aisin "have grown to be comparable in size to Toyota and are technically autonomous. They can design their own subsystems and components and have complete prototype and test capabilities. They are involved in the earliest concept states at Toyota, and they often develop sketches before Toyota has created a contract or even developed formal specifications for the subsystem" [Morgan and Liker, 2006, p. 183].

SUGGESTED IMPLEMENTATION: Modern knowledge about lean supply chain management is taught at a master's level in select top business schools. Every successful enterprise should strive to acquire and bring the knowledge in-house, either by sending their own managers to relevant educational programs or by hiring outside professionals. Benchmarking with the most progressive corporations will aid in determining gaps in practice and needed strategic improvements. For competitive reasons, strategies should be implemented as fast as possible. It is recommended that only the business schools that have solid knowledge about Lean supply chains be used.

LAGGING FACTORS: Traditional relationships between buyers and suppliers, manifested by all the bad habits: *over-the-wall* communications, lack of trust, exploitation, short term goals, buying from the lowest bid, unstable specifications, wars about quality and timeliness, dual inspections, requirements for certifications of individual parts, and other issues.

SUGGESTED READINGS:
- J.K. Liker, *The Toyota Way: 14 Management Principles from the World's Greatest Manufacturer*. McGraw-Hill Professional, 2004.
- J.M. Morgan and J.K. Liker, *The Toyota Product Development System*. New York: Productivity Press, 2006.
- K. Bozdogan, *Supplier Networks Transformation Toolset*, Lean Advancement Initiative, MIT (http://lean.mit.edu), 2004.

LEAN PRINCIPLE 3: Flow

ENABLER: 4. Systems Engineers to accept Responsibility for Coordination of PD Activities.

SUBENABLER(S) AND USE RANKINGS:
1. **Promote maximum seamless teaming of Systems Engineers and other Product Development (PD) engineers. (U = 0.36)**
2. **SE to regard all other engineers as their partners and internal customers, and vice versa. (U = 0.12)**

VALUE PROMOTED: Teaming of SE and other PD engineers, better planning and execution of programs

WASTE PREVENTED: All categories of waste.

EXPLANATION: Ideal Systems Engineering (SE) is supposed to function like a healthy nervous system in the body: well-integrated with all other body parts and across their interfaces, sensing all states in real time, coordinating all activities, and providing real-time signals, inputs, outputs and feedback between parts, subsystems and interfaces. Good SE identifies pain (issues) early and initiates immediate mitigation activities. In order to have such an ideal SE, excellent teaming is needed between SE engineers and other Product Development (PD) engineers. Good teaming requires that both Systems Engineers and other engineers see their primary responsibility as serving the needs of customers, their programs, and other engineers and managers.

SUGGESTED IMPLEMENTATION:
- Chief Engineer (or equivalent role) should promote and demand good teaming between SE and other engineers.
- Ensure all roles understand their contributions to the product development.
- A short training for all engineers (should be conducted at the corporate level because it serves all programs), demonstrating examples of good and bad teaming, and impact on programs.
- Lots of mentoring and good examples.

LAGGING FACTORS:
- Separate and bureaucratic Systems Engineering function that is poorly integrated with technical engineering processes, and focused only on production of artifacts.
- Lack of good communications, trust, openness, honesty, respect.

SUGGESTED READING:
- J.M. Morgan and J.K. Liker, *The Toyota Product Development System*. New York: Productivity Press, 2006.

LEAN PRINCIPLE 3: Flow

ENABLER: 4. Systems Engineers to Accept Responsibility for Coordination of PD Activities.

SUBENABLER(S) AND USE RANKINGS:
3. **Maintain team continuity between phases to maximize experiential learning. (U = 0.04)**
4. **Plan for maximum continuity of Systems Engineering staff during the program. (U = 0.20)**

VALUE PROMOTED: Smooth value flow between program phases.

WASTE PREVENTED: Information loss caused by changes in staff and by program discontinuities.

EXPLANATION: Some complex technology programs extend over many years, in exceptional cases reaching 10 to 15 years, or more. Long programs may suffer from funding instability. Such long time frames cause employees to retire, change jobs, and new ones to be hired. Each change of experienced personnel introduces negative impacts on the program, such as learning errors, *reinventing the wheel*, changes in approaches, *requirements creep*, and loss in organizational memory. For these reasons, shorter programs are definitely more conducive to efficiency and stability.

Subenabler 3.4.3 addresses the potential loss of experiential learning that occurs when transitioning across program phases and milestones. Ideally, in order to minimize the knowledge loss that occurs between different program phases, the same team should be present across each phase. Good standardization and configuration management aid the continuity.

In order to minimize the negative impact of long programs, subenabler 3.4.4 promotes maximum continuity of SE staff. In practice, enterprise management should provide incentives for experienced staff to stay with the program, or at least be available to return to when needed. Good capture of program and corporate memory is helpful, but must be balanced against excessive bureaucracy.

SUGGESTED IMPLEMENTATION:
- Create corporate procedures for continuity of critical personnel, for backup personnel for key stakeholders, and seamless configuration management and program memory across program discontinuities.
- Build a robust body of knowledge, organized by program, to ensure transfer of knowledge if the team is no longer available. Create a searchable database with lessons learned and best practices.
- Budget realistically for the fulfillment of the procedures.

LAGGING FACTORS:

- Long programs.
- Program funding instability.
- Staff changes in long programs.
- Poor configuration management and poor databases.

SUGGESTED READING: N.A.

LEAN PRINCIPLE 3: Flow

ENABLER: 5. Use Efficient and Effective Communication and Coordination.

SUBENABLER(S) AND USE RANKINGS:
1. Capture and absorb lessons learned from almost all programs: "never enough coordination and communication." (U = −0.52)

VALUE PROMOTED: Good communications are critical to predictable and robust execution of work flow and value delivery.

WASTE PREVENTED: All categories of waste.

EXPLANATION: Planning and training for efficient and effective communications are one of the most critical aspects of any Lean organization. Neither too little nor too much communication is good for a program. Too little communication (notorious in large programs) may cause a large variety of problems, for example: rework of uncoordinated tasks, issues remaining in hiding and growing to crisis proportions, uncoordinated interfaces, lack of synchronization of activities, non-optimum solutions, and many others. Too much spontaneous communication is also undesired: Complex programs involve thousands of stakeholders, including suppliers. The theoretical number of potential paths for information to flow between n people is $n*(n-1)/2$. With 1,000 people, the number is 499,500! So ad-hoc communications "of everybody with everybody else," including unneeded emails, can consume precious program time. The amount of communication and coordination must be just right. This requires experience, good planning and focused training.

Recent problems with the educational system contribute to the problem. Most young engineers lack effective communication skills. The K-12 system is less than perfect in the United States and many other countries. And engineering education tends to focus on analysis, design, and laboratory skills, with almost no attention paid to coordination and communications. Many young engineers prefer "talking" to their computers in their own cubicles and are not aware of the overwhelming need for efficient and effective communications with other human beings in complex programs. Managers hiring new employees must be aware of this widespread problem and plan accordingly.

In short, communications and coordination must be well planned at the beginning of a program, and the entire team of stakeholders must be well trained in such practice, or the program may be destined for trouble.

Some managers tend to view automated computer tools as a substitute for good personal communications. While a well-selected tool may indeed help in capturing and sharing information, it will never be a substitute for intelligent, effective, and efficient human-to-human communication using the broadest possible bandwidth. For difficult communications involving brainstorming, negotiations, and consensus building, nothing is better than face-to-face co-located conversation. Virtual communication is not quite the same thing!

A number of enablers deal with communications and coordination. Program stakeholders are urged to implement all of them to the best of their ability.

Morgan and Liker [2006, p. 261] present this Lean view of Toyota communications:

- If everyone is responsible, no one is responsible.

- If everyone must understand everything, no one will understand anything very deeply.

- If all communication is going to everyone, no one will focus on the most critical communication for their [sic] role and responsibility.

- If you inundate your people with reams of data, no one will read it.

SUGGESTED IMPLEMENTATION: An experienced top-level manager should issue a comprehensive memo or standard for efficient and effective communications and coordination in an enterprise or program: who with whom and when and how. Training materials should be prepared, and all stakeholders in the program should complete such training. The recommended training scope is one to two hours, plus frequent subsequent mentoring, a lot of good examples, and disciplined practice. Since all programs in a corporation benefit from this training, it should be organized and performed at the corporate level for all employees, soon after hiring. At a minimum, the rules should include the following points:

- Send your email only to those who truly need to read it.

- Instead of verbose text, send only a short email attaching a standardized A3 form presenting an issue.

- If you have a question, immediately locate the person who can answer it and seek the answer without wasting precious program time.

- Unless the task is routine, identify the internal customer (by name) and coordinate the task details with the person before starting work, so that you can execute the task *right the first time*.

- Let the appropriate managers know immediately if you identify a risk to a program, delay, or a cross-functional issue, using a standard A3 form.

LAGGING FACTORS:
- Culture of ad-hoc communications, lack of planning for good communications, and lack of training in effective communications.
- Culture of isolated engineers working at their computers with all communications via electronic means.
- Fear-inducing managers who blame employees rather than help and mentor them.

SUGGESTED READING:
- J.K. Liker, *The Toyota Way: 14 Management Principles from the World's Greatest Manufacturer*. McGraw-Hill Professional, 2004.
- E.M. Murman, *Lean Aerospace Engineering*, Littlewood Lecture AIAA-2008-4, January 2008.
- J.M. Morgan and J.K. Liker, *The Toyota Product Development System*. New York: Productivity Press, 2006.
- J.H. Gittell, *The Southwest Airlines Way*. McGraw Hill, New York, 2003.

LEAN PRINCIPLE 3: Flow

ENABLER: 5. Use Efficient and Effective Communication and Coordination.

SUBENABLER(S) AND USE RANKINGS:
2. Maximize coordination of effort and flow (one of the main responsibilities of Lean SE). (U = 0.24)

VALUE PROMOTED: Predictable and robust flow of work.

WASTE PREVENTED: Rework, delays, frustrations.

EXPLANATION: This subenabler emphasizes one of the most important responsibilities of Systems Engineers, namely the proper planning and coordination of effort and flow. For the flow of work to proceed robustly, predictably, *right the first time*, the effort of each task in the flow should be predictable, well planned, and coordinated with other tasks and with the overall flow. This requires common planning between IPTs, functions, and major suppliers. Co-located Value Stream Mapping is an excellent tool for such planning. Planning for maximum concurrency minimizes schedules. Load should be leveled as much as possible. The critical path effort should be clearly identified. (Time buffers may be used to compensate for task arrival variability.) Task expectations should be defined and well understood. Tasks should do no more and no less than what is needed to create value to the customer. When effort is coordinated as described, value stream is understood, and dependencies and risks are identified, then flow can proceed naturally. The role of SE is to monitor the state of project to ensure continuous and predictable flow of value-added effort.

Even the best plan requires some flexible adjustments to tasks. Therefore, during program execution, Systems Engineers must track and coordinate tasks in real time, proactively, to assure good progress.

SUGGESTED IMPLEMENTATION:
- Solid understanding of the need to coordinate effort and flow on the part of Systems Engineers.
- Training of SE engineers and managers in Value Stream Mapping and Lean flow characteristics.
- Co-location of key SE and functional managers and engineers in the *war room* during the planning phase is highly recommended.
- Disciplined adherence to the Value-Stream Map.
- Mentoring, good examples.

LAGGING FACTORS:

- Ad-hoc rushed planning.
- Lack of culture of precise coordination of effort between IPTs and functions.
- Lack of Lean knowledge.
- *Stovepipe* culture.

SUGGESTED READING:

- J.M. Morgan and J.K. Liker, *The Toyota Product Development System*. New York: Productivity Press, 2006.

LEAN PRINCIPLE 3: Flow

ENABLER: 5. Use Efficient and Effective Communication and Coordination.

SUBENABLER(S) AND USE RANKINGS:
3. **Maintain counterparts with active working relationships throughout the enterprise to facilitate efficient communication and coordination among different parts of the enterprise and with suppliers. (U = 0.50)**
4. **Use frequent, timely, open, and honest communication. (U = 0.48)**
5. **Promote direct informal communications immediately as needed. (U = 0.76)**

VALUE PROMOTED: Open, honest and direct communications conducted immediately between the parties involved, as needed, without waiting or bureaucracy.

WASTE PREVENTED: Primarily waiting waste, but other categories as well.

EXPLANATION: Consider two programs: one heavily *stovepiped*, the other one Lean. In the former, when an engineer needs to ask a question of his or her counterpart in a different department, he or she may be instructed to go only through the chain of command, so transmits the question to a direct manager. The manager, possibly busy, pushes the question onto his or her colleague manager or superior with some delay. That person, again, after some possible delay, passes the question onto another subordinate. The answer travels the same way in reverse. Typically, each handoff introduces information noise and takes away from clarity. The final answer may reach the originator hours, days, or even weeks later, losing information quality in all the handoffs. In contrast, in a Lean program the originator knows his or her counterpart in other departments, is empowered to contact that person immediately, discuss the issue until both are satisfied, and the communication is closed in minutes. If this example is multiplied by the number of employees involved in such communications, we see how the resultant waste can take up a significant portion of program schedule and budget.

In Lean organizations, engineers should be empowered and instructed to communicate directly and informally in real time with their counterparts in other departments with whom they are expected to communicate. This minimizes the number of handoffs (each handoff is said to waste 50% of information) and saves precious program time. In large programs, the number of employees is so large that just finding the right person to contact takes valuable time. [Recall: The theoretical number of potential paths for information to flow between n employees is $n*(n-1)/2$]. To overcome such handoff and waiting wastes, experienced managers encourage engineers to set up and maintain counterparts throughout the enterprise to facilitate efficient communication and coordination. Each pair of engineers knows each other's emails, phones, and locations, and communicates immediately as needed, directly, without having to go through managers or immense bureaucratic hurdles. Direct links also promote teamwork and collegiality.

The same principle applies even more powerfully to the pairing of engineers between a buyer and suppliers. In this case, however, before a pair can be empowered to communicate directly, engineers must complete a short training on the legality and limitations of the communications in order to prevent unintended *requirements creep* (see subenabler 3.5.9). In this case, well-connected counterparts can save significant time and effort clarifying requirements and coordinating details directly, rather than going through a bureaucratic maze—as is practiced in traditional *over-the-wall* relationships with suppliers.

Subenablers 3.5.4 and 3.5.5 emphasize the need for such communications to be open and honest and to be initiated immediately as needed.

SUGGESTED IMPLEMENTATION: Prepare a short training session on the policy of empowerment and encouragement to communicate with counterparts directly, immediately, openly, and honestly. Include in training the points about limitations and legal aspects of communications with suppliers. Administer such training to all engineers. Mentor in these practices and offer good examples.

Since communication culture benefits all programs in a corporation, it should be administered at the corporate level to all employees.

LAGGING FACTORS:
- *Stovepipe* organization, managers demanding that "everything goes through me."
- Culture of fear.
- Lack of adequate knowledge of Lean practices.

SUGGESTED READING:
- J.K. Liker, *The Toyota Way: 14 Management Principles from the World's Greatest Manufacturer*. McGraw-Hill Professional, 2004.
- J.M. Morgan and J.K. Liker, *The Toyota Product Development System*. New York: Productivity Press, 2006.

LEAN PRINCIPLE 3: Flow

ENABLER: 5. Use Efficient and Effective Communication and Coordination.

SUBENABLER(S) AND USE RANKINGS:

6. **Use concise one-page electronic forms (e.g., Toyota's A3 form) rather than verbose unstructured memos to communicate and keep detailed working data as backup. (U = −0.28)**

7. **Report cross-functional issues to be resolved on concise standard one-page forms to Chief's office in real time for his or her prompt resolution. (U = −0.33)**

VALUE PROMOTED: Disciplined, efficient, and effective written communications.

WASTE PREVENTED: Directly, the time wasted for inefficient communications. Indirectly, all categories of waste are reduced.

EXPLANATION: Today's engineers are not known for their great writing skills. Instead of forcing engineers to waste precious program time struggling to write verbose emails, memos, or reports, which often end up incomprehensible anyway, Lean organizations practice written communications on pre-formatted sheets of A3 size paper. (Why A3? Historically, this was the largest paper that could fit through a fax machine.) Programs should design their own forms. (Examples are available in the literature listed below and on the Internet.) For example, the form intended to report a cross-functional issue might have the following fields: author (name, email, location, phone, position), date, issue title, brief explanation, risk impact, stakeholders affected, root cause analysis, suggested alternative solutions, proposed solution, and possibly more advanced data such as effort, cost, and schedule impact. Engineers are formally trained to fill these forms out efficiently and promptly and instructed to send them as an email attachment to only tightly selected individuals, without excessive use of the email "cc:" function. Not every field needs to be filled every time: the priority is to signal an issue immediately with minimum bureaucracy. No other attachments, and particularly no verbose reports, should be sent in any first such communication unless explicitly requested by the recipient or another stakeholder. Typically, the form should be sent to the office of Chief SE for followup decisions or actions.

These forms are easy to create, read, and organize into agendas for *integrative events*, to sort and classify as a risk item, and to file systematically. Once employees become comfortable using these forms, their reading perception improves. Companies that use A3 forms realize significant savings in time and frustrations. In contrast, verbose unstructured emails and memos consume precious program time and are a source of never-ending frustrations.

Toyota has perfected the use of such A3 forms. They represent not only a
paper form, but also a highly disciplined process of expressing complex
thoughts accurately and efficiently on a standardized single pre-formatted
sheet of paper.

An Internet search on "A3 form" hits a large number of useful samples.

Consider the following quotes regarding A3 reports [Morgan and Liker, 2006,
p. 269]:

- "Force yourself to filter and refine your thoughts to fit one sheet of paper in
  such a way that management has all of the [needed information] by reading
  a single piece of paper—it is the essence of lean."

- "A3 is much more about disciplined thinking than it is about any particular
  writing technique."

- "A well-socialized Toyota engineer almost intuitively knows that getting
  input from all the right people is necessary and he or she has learned from
  experience the value of presenting information concisely and visually.
  Without this cultural context, the A3 is a mechanical requirement, an
  exercise in summarizing complex information to please the boss. Without
  deep thought and consensus-based process, the A3 report is a tool generated
  for the wrong reason. In lean thinking, it is the process of producing not the
  result of producing the A3 that makes it a powerful method."

- A problem-solving A3 would succinctly state the problem, document the
  current situation, determine the root cause, suggest alternative solutions,
  suggest the recommended solution, and have a cost-benefit analysis" [Liker,
  2004, p. 157].

SUGGESTED IMPLEMENTATION:
- Explore the numerous examples of A3 forms available on the Internet, select
  the most appropriate, and redesign to fit company or organization needs.
- Train all engineers, technicians, and managers in the proper use of the forms
- Set up databases to capture forms efficiently and create easy searches and
  access.
- Mentor, offer good examples, and disallow verbose ad-hoc texts.

LAGGING FACTORS:
- Ad-hoc communication culture.
- Tolerance for imperfect verbose texts.
- Lack of policies for efficient communications.

SUGGESTED READING:
- J.K. Liker, *The Toyota Way: 14 Management Principles from the World's
  Greatest Manufacturer*. McGraw-Hill Professional, 2004.
- J.M. Morgan and J.K. Liker, *The Toyota Product Development System*. New
  York: Productivity Press, 2006.
- http://www.coe.montana.edu/ie/faculty/sobek/a3/index.htm, authored by
  Duward Sobek.

LEAN PRINCIPLE 3: Flow
ENABLER: 5. Use Efficient and Effective Communication and Coordination.
SUBENABLER(S) AND USE RANKINGS: 8. **Communicate all expectations to suppliers with crystal clarity, including the context and need, and all procedures and expectations for acceptance tests and ensure the requirements are stable. (U = 0.35)**   9. **Trust engineers to communicate with suppliers' engineers directly for efficient clarification, within a framework of rules (but watch for high-risk items that must be handled at the top level). (U = 0.36)**
VALUE PROMOTED: *Right the first time* deliveries of quality parts and supplied subsystems. Predictable and robust testing. Verification and qualification processes. Program cost and time savings.
WASTE PREVENTED: Defects, delays, and related managerial and legal costs of resolving disputes on non-conforming parts.
EXPLANATION: Modern programs outsource 60–95% or more of value creation to suppliers; therefore, good relations with suppliers is critical. Yet, traditional relationships, still frequently practiced, are notorious for conflicts over cost, timeliness, and quality of supplies. A frequent scenario proceeds as follows. A buyer submits rushed, late, and incomplete specifications to a supplier. The specifications lack clarity and precision regarding tests that the supplied items must pass. The specifications fail to include a broader description of need and context, a feature often necessary to help suppliers do the job *right the first time*. The channels for easy clarification of technical information do not exist or are burdensome and bureaucratic. The supplier delivers the ordered part on time and within the agreed price only to fail some receiving tests that were not explicitly included in the original specifications. The buyer insists that the tests are legitimate. The supplier points to the lack of clear test descriptions in the specifications. At this time, the *big guns* enter the picture: contract lawyers, high-level managers. After a significant amount of negotiations, frustrations, and delays, if not additional payments, a forced agreement is reached to rework or retest the failed part. In the meantime, precious program time and budget have suffered a hit. The transaction leaves all parties unsatisfied, and the supplier is removed from the preferred list.

Who is at fault? Of course, the buyer. The buyer has both a legal and professional responsibility to:

1. Provide suppliers with timely and crystal clear specifications not only for parts, but also for all tests that the parts are expected to pass. The supplier can then make certain the parts pass all necessary tests before delivery. Tests can be witnessed by the buyer—then no second test should be needed at the receiving end (the second test represents a waste of rework and waiting). In ideal buyer-supplier relationships based on long-term partnerships, supplier quality will be perfect, and buyers will trust their suppliers. This ideal is achievable but takes work.

2. Ensure all specifications passed to suppliers are stable.

3. Enable efficient and effective direct channels for real-time technical clarifications of requirements. Clarifications must occur within a framework of rules. Engineers empowered to clarify must first complete solid training in such rules to prevent *requirements creep*. Exception: High-risk items must be handled at the top level.

4. Strive to develop long-term seamless relationship based on partnership, honesty, openness, sharing of data, and fairness.

Nothing is obsolete in the mantra by Deming [1982, p. 433]: "Successful cooperation with suppliers of parts, especially of critical parts, and success in tests and adjustments of subassemblies, reduce to a rarity any major trouble in tests of the final assembly" (Deming, 1982, p. 433).

Toyota is well regarded for its excellent supply network. Four quotes illustrate the network.

- Toyota upholds the idea that "it is more expensive in the long run to pick the cheapest supplier if this supplier is not ready to meet your requirements." [Morgan and Liker, 2006, p. 192]

- The underlying purpose of developing seamless partnerships with suppliers and the product development team is that the customer cannot tell the difference between the two. "Toyota recognizes this and makes sure that every car part reflects Toyota quality. To achieve this, they make [sic] every supplier an extension of Toyota's PD process and lean logistics chain. Toyota delegates tasks to suppliers, but ultimately, it is Toyota that is fully responsible and fully accountable for all subsystems as well as the final product. Outsourcing does not absolve Toyota of responsibility and accountability." [Morgan and Liker, 2006, 180]

- "When Toyota invites a supplier to send guest engineers, it is a significant commitment to long-term co-prosperity. The supplier knows they have [sic] earned a long-term place in the Toyota enterprise. These are highly coveted positions. The supplier will become intimately familiar with Toyota's product development practices and get advanced information on new model programs". .... The suppliers know that their engineers benefit a great deal from their experience at Toyota, learning 'The Toyota Way' of engineering. [Morgan and Liker, 2006, pp. 193–194]

- "New U.S. suppliers seldom meet Toyota's expectations at first, and it can be challenging for both parties to work together. For example, ... the Toyota designer developed a supplier by teaching the proper way to collect accurate tolerance data, analyze it, and develop corrective actions to improve quality in manufacturing. In the traditional supplier-company relationship, this seldom happens" [Morgan and Liker, 2006, p. 191].

SUGGESTED IMPLEMENTATION: Transitioning from traditional *over the wall* bureaucratic relationships with suppliers and selecting suppliers based on lowest bid to modern supply-chain management based on partnership and commonality of long-term business goals is not easy. Effective implementation includes the following basic steps:
- Top managers of a Purchasing Department (or equivalent name) should obtain excellent education and knowledge in modern Lean supply-chain management or hire such managers.
- The managers should comprehensively implement new strategic management of supply chains, if necessary with assistance from expert consultants. This can be a multi-year effort. In fact, such an effort never ends because new suppliers join the network all the time.
- Train other stakeholders (engineers and managers authorized to interact with suppliers directly) in the applicable rules to build common culture of cooperation, teamwork, and partnership, without risking unauthorized *requirements creep* or contract change.
- Patiently keep developing relations with suppliers that are based on openness, teamwork, good communications, mutual interest, fairness, long-term partnership, but demand Lean knowledge and practice of all suppliers.

LAGGING FACTORS:
- Traditional selection of suppliers from lowest bid for a single contract.
- Multiple suppliers for same items.
- *Over the wall* distrustful relationships with suppliers.
- Tradition of providing suppliers with poor or incomplete specifications and being late.
- Not telling suppliers how parts must be tested.

SUGGESTED READING:
- J.M. Morgan and J.K. Liker, *The Toyota Product Development System*. New York: Productivity Press, 2006.
- E.M. Murman, *Lean Aerospace Engineering*, Littlewood Lecture AIAA-2008-4, January 2008.
- K. Bozdogan, *Supplier Networks Transformation Toolset*, Lean Advancement Initiative, MIT (http://lean.mit.edu), 2004.
- W.E. Deming, *Out of the Crisis*. Massachusetts Institute of Technology, Center for Advanced Engineering Study, 1982.

LEAN PRINCIPLE 3: Flow
ENABLER: 6. Promote Smooth SE Flow.

SUBENABLER(S) AND USE RANKINGS:
1. Use formal frequent comprehensive *integrative events* in addition to programmatic reviews. (U = 0.00):
   a. Question everything with multiple "whys." (U = −0.04)
   b. Align process flow to decision flow. (U = 0.16)
   c. Resolve all issues as they occur in frequent *integrative events*. (U = −0.08)
   d. Discuss tradeoffs and options. (U = 0.72)

VALUE PROMOTED: Steady, predictable, and robust flow of PD program
WASTE PREVENTED: All waste categories, including schedule and budget overruns.
EXPLANATION: In Chapter 4, Section 4.2 contains a description of the process termed "Lean Product Development Flow (LPDF)" [Oppenheim, 2004]. This process is intended for a special class of smaller and low-risk projects. But, certain process practices apply to all programs, and they are summarized here. This process promotes frequent and comprehensive *integrative events* (meetings) because long time intervals between reviews are the PD equivalent of manufacturing inventory, which is a source of significant waste. In PD, issues should be identified and dealt with in real time when they are small and easy to mitigate. Concealed or left unaddressed, issues tend to grow, even reaching crisis proportions. If small issues can be addressed and resolved during daily standup meetings, so much the better. However, many problems, especially cross-functional issues and larger risks, require a more comprehensive coordination between functions and resolution by top managers, often requiring longer meetings. These meetings should be held frequently and periodically. Weekly meetings are recommended. All new significant issues should be addressed and mitigated or resolved during the meetings. In addition, the meetings serve to perform precise planning for next work period and to address all new questions. *Integrative events* should have well-scripted agendas. The Chief Engineer (or equivalent role) or his or her deputy should conduct the meetings, asking multiple questions "why" to stimulate better decision making, better alignment of process and decision flow, and better discussion of current tradeoffs and options. Many decisions can be made on the spot. Experience indicates that best programs allocate enough time for such comprehensive meetings, provided the meetings are planned and executed productively. Meeting agendas should be based on those issues and questions submitted by program stakeholders in real time during the preceding week (preferably on A3 forms).

SUGGESTED IMPLEMENTATION:
- The implementation should start with solid training (about one day) for all technical and managerial employees in the program. Training should cover benefits and scope of periodic *integrative events*.
- The program should be led by a Chief SE who possesses the following competences: domain, Systems Engineering, Lean, and leadership. If such leaders are lacking, companies should begin to groom them ASAP, as described in Oppenheim [2004].

LAGGING FACTORS:
- Long periods between *integrative events*.
- Culture of only short bureaucratic status meetings; lack of frequent comprehensive *integrative events*.
- Focus on bureaucratic artifact production rather than on immediate resolution of issues.
- Dissolved management and lack of leadership.
- Lack of good planning.

SUGGESTED READING:
- B.W. Oppenheim, Lean product development flow. *Journal of SE*, 2004, **7**, 4, pp. 352–376.
- J.M. Morgan and J.K. Liker, *The Toyota Product Development System*. New York: Productivity Press, 2006.

LEAN PRINCIPLE 3: Flow

ENABLER: 6. Promote Smooth SE Flow.

SUBENABLER(S) AND USE RANKINGS:
2. Be willing to challenge the customer's assumptions on technical and meritocratic grounds and to maximize program stability, relying on technical expertise. (U = 0.48)

VALUE PROMOTED: Better requirements. More direct path to the validation of the value proposition

WASTE PREVENTED: Starting the program with poor requirements leads to dysfunctional and failed programs. All categories of waste are affected.

EXPLANATION: In an ideal world, value proposition should be perfectly captured in requirements, specifications, goals, Concept of Operations (CONOPS), and other such documents. Yet, many acquisition programs have the funding approved and proceed to a Request for Proposals (RFP) phase with incomplete, incorrect, or mutually conflicted top-level requirements. Numerous reasons contribute to this: too many stakeholders, lack of expertise (or even competence) in the product or mission on the part of customer stakeholders, rushed jobs by government agencies pressed by funding urgency, lack of coordination of value with actual end users, rotation of government employees in the middle of value formulation, assumptions by customers that some aspects of value are self-evident and do not need to be spelled out, and many others. Deming [1982, p. 143] points out that "the customer's specifications are often far tighter than he needs. It would be interesting to ask a customer how he arrives at his specifications, and why he needs [what] he specifies."
Recent U.S. government publications (see Section 5.2) indicate that practically all governmental programs suffer from some of these deficiencies. Therefore, contractors must be willing to challenge customer assumptions when doing so is justified for the good of the customer and for value. The challenge must be based on technical and cost/schedule merits, not on arbitrary preference of the contractor. Changes suggested must be justified by experts.

SUGGESTED IMPLEMENTATION: This subenabler is relatively easy to implement; all that is needed are instructions from Chief SE to the program team whereby, "if you see something wrong with the requirements, including customer's requirements, provide a solid technical justification for what and why it is wrong, and report it up the chain of command to the Program Office" (the sole party authorized to challenge customers).

LAGGING FACTORS:

- Acceptance of everything that comes from customers as *sacred*, not to be challenged under any circumstances, even if it their requirements are wrong or would increase program scope.
- Rigid mental walls between the customer and the contractor stakeholders.
- Bureaucratic management.

SUGGESTED READING:

- G.H. Plenert, *Reinventing Lean: Introducing Lean Management into the Supply Chain*. Butterworth-Heinemann, 2007.
- J.M. Morgan and J.K. Liker, *The Toyota Product Development System*. New York: Productivity Press, 2006.
- D. Lempia, *Using Lean Principles and MBE in Design and Development of Avionics Equipment at Rockwell Collins*, 26th *International Congress of the Aeronautical Sciences,* Anchorage, AK, September 16, 2008, paper 6.7.3.
- A.C. Haggerty and E.M. Murman, *Evidence of Lean Engineering in Aircraft Programs*, 25th *International Congress of the Aeronautical Sciences,* Hamburg, Germany, September 2006.
- W.E. Deming, *Out of the Crisis*. Massachusetts Institute of Technology, Center for Advanced Engineering Study, 1982.

LEAN PRINCIPLE 3: Flow

ENABLER: 6. Promote Smooth SE Flow.

SUBENABLER(S) AND USE RANKINGS:
3. Minimize handoffs to avoid rework. (U = −0.04)

VALUE PROMOTED: *Right the first time* execution of tasks.

WASTE PREVENTED: Rework, delays, frustrations.

EXPLANATION: As explained under subenablers 3.5.3–3.5.5, information handoffs represent nasty waste in organizations that create and move around lots of information. The well-known children's game "telephone" introduces this succinctly. Consider the mechanism of a handoff in a complex work environment: The person asking a question has in his or her mind the rich understanding of a problem in its context. Based on this rich context, he or she formulates a question involving only a few words in a natural language, each word having several dictionary meanings. The recipient hears only the imperfect question and not the background context. So, this person answers accordingly, also using inherent linguistic ambiguities. Each verbal handoff is said to lose 50% of the information intended. If performed in series n times, the end product contains $(0.5)^n$ of the original meaning. Thus, a not-infrequent practice of following the chain of command and directing a question orally via this chain—from employee to a manager to another employee, and back—would transmit only 6.25% of the relevant information. In addition, people may not be available to answer the question immediately, and precious program time is wasted waiting for answers. Of course, written questions survive handoffs better than spoken, but still suffer from context losses. As Morgan and Liker [2006, 73] point out: "In manufacturing, [. . .handoff] means moving parts and products unnecessarily. In product development, it means unnecessary handoffs from one overly specialized activity to another, and more specifically, information changing hands, whether by word, picture, or data exchange. This waste leads to the loss of momentum, information, and accountability in the process. It is also commonly accepted dysfunction in traditional PD systems."
In contrast, in lean organizations employees are empowered, trained, and directed to locate needed sources of information for direct and immediate contact.

SUGGESTED IMPLEMENTATION: Empowering the employees to coordinate questions directly requires a somewhat advanced lean culture based on trust, honesty, openness, and respect and also responsibility and accountability. Thus, these aspects of organizational culture should be implemented as soon as possible. Experience indicates that transition from traditional silo culture to Lean culture takes time and effort and must be driven by top managers in a systematic way, starting with a comprehensive training in Lean.

LAGGING FACTORS:
- Silo-type organizations.
- Lack of trust in employees being able to communicate directly.
- Lack of training in effective communications.

SUGGESTED READING:
- J.M. Morgan and J.K. Liker, *The Toyota Product Development System*. New York: Productivity Press, 2006.
- D. Lempia, *Using Lean Principles and MBE in Design and Development of Avionics Equipment at Rockwell Collins*, 26th *International Congress of the Aeronautical Sciences,* Anchorage, AK, September 16, 2008, paper 6.7.3.

LEAN PRINCIPLE 3: Flow

ENABLER: 6. Promote Smooth SE Flow.

SUBENABLER(S) AND USE RANKINGS:

4. Optimize human resources when allocating VA and RNVA tasks.
   (U = 0.08):

    a. Use engineers to do VA engineering. (U = 0.36)

    b. When engineers are not absolutely required, use non-engineers to do RNVA (administration, project management, coasting, metrics, program, etc.) (U = 0.08)

VALUE PROMOTED: Better utilization of human resources, program cost savings.

WASTE PREVENTED: Waste of expensive human resources. Frustrations from engineers who are asked to perform non-engineering work.

EXPLANATION: Many programs are burdened with administrative tasks that represent Required Non-Value-Added (RNVA) activities. A not-infrequent practice is to utilize employees for administrative RNVA tasks, regardless of their job description and professional preparation. This subenabler promotes utilizing engineers for VA engineering work, and assigning simple administrative (bureaucratic?) RNVA tasks to less-expensive administrative employees. Even when an engineer appears to have nothing to do, instead of burdening him or her with RNVA tasks, it is often cheaper in the long term to ask a non-engineer to do the RNVA task and to ask the engineer to focus on continuous improvement, learn new knowledge, upgrade checklists and procedures, implement lessons learned, observe and learn from another departments or suppliers, or other such tasks.

Clearly, there may be exceptions to this rule: periods of emergencies, *all-together* work marathons to meet deadlines, for example, but they should be called for judiciously and only rarely.

SUGGESTED IMPLEMENTATION: Easy to implement. A simple command from the Program Office to use only non-engineers for administrative tasks and to use engineers for actually needed engineering tasks or for continuous improvement, training, rotation, etc. A RASI tool is effective for assigning appropriate tasks to appropriate stakeholders.

LAGGING FACTORS: Managers who wish to employ everybody to 100% of capacity all the time and do not appreciate the value of continuous improvement, continued education, and professional activities that pay long-term dividends.

SUGGESTED READING:
- J.M. Morgan and J.K. Liker, *The Toyota Product Development System*. New York: Productivity Press, 2006.

LEAN PRINCIPLE 3: Flow

ENABLER: 6. Promote Smooth SE Flow.

SUBENABLER(S) AND USE RANKINGS:
5. Ensure the use of the same measurement standards and database commonality. (U = 0.13)

VALUE PROMOTED: *Right the first time* flow of design and manufacturing.

WASTE PREVENTED: Rework, information churning, the high cost of using 2D drawings, overprocessing, and, indirectly, all other types of waste.

EXPLANATION: At its simplest, this subenabler promotes making sure that the same set of units are used within an entire enterprise to avoid disasters such as the infamous Mars Climate Orbiter failure caused by different stakeholders using different unit systems. Significant savings can be realized when all stakeholders use the same measurement standards and a common central database, based on a common and well-integrated 3D tool set. The modern 3D tools are so advanced that the entire system geometry and all characteristics, dimensions, material properties, notes, manufacturing instructions, tool paths, and assembly simulations can now be handled efficiently within an integrated set of CAE tools, all using a common database. Ideally, an integrated tool set should include 3D modeling and a complete set of analyses, assembly simulation and tolerance checking, manufacturing tool paths design, and associated design and manufacturing functions. The database information should be divided into different files or levels each with distinct and secure access to be used by specific suppliers who, as a condition for a contract, must install compatible workstations to read (but not alter!) 3D part definitions. This approach saves significant costs by totally eliminating the need for 2D drawings, a costly and error-prone dinosaur of the past.
Sharing a common database among all stakeholders facilitates the use of common measurement standards, common reference planes and tolerance standards, and other manufacturing standards.

SUGGESTED IMPLEMENTATION:
- In every contract spell out the system of units to be used and disseminate this information to all stakeholders, with compliance feedback.
- Insist that 2D drawings be abandoned throughout all programs.
- Purchase a well-integrated 3D modeling and analysis tool set with options supporting an entire PD lifecycle. Demand that all suppliers of parts install reading stations. Then give suppliers controlled electronic access to only the data they need.
- Construct and use a well–thought-out central database and central release system.
- Train all engineers in the use of the tools.

These steps are already implemented at most high-tech companies. The steps should be migrated to lower tier suppliers.

LAGGING FACTORS:

- The use of 2D drawings. This is often required in government contracts by inexperienced staff who are not aware of progress and the huge benefits of using 3D integrated toolset.
- Government contracts, and *earned-value* metrics that measure progress by the number of drawings released. This is a 19th-century requirement applied in the 21st century.
- Mergers of companies with different tools, databases, and standards.
- Resistance from traditional suppliers.

SUGGESTED READING:

- A. Haggerty, *Lean Engineering, lecture*, MIT Minta Martin lecture series (available on DVD from the MIT LAI).
- H. McManus, A. Haggerty, and E. Murman, Lean engineering: a framework for doing the right job right, *The Aeronautical Journal*, February 2007, **111**, 1116, pp. 105–114.
- A.C. Haggerty and E.M. Murman, *Evidence of Lean Engineering in Aircraft Programs*, 25th *International Congress of the Aeronautical Sciences*, Hamburg, Germany, September 2006.

LEAN PRINCIPLE 3: Flow

ENABLER: 6. Promote Smooth SE Flow.

SUBENABLER(S) AND USE RANKINGS:
6. Ensure that both data deliverers and receivers understand the mutual needs and expectations. (U = 0.36)

VALUE PROMOTED: *Right the first time* execution of tasks, reduction of program schedule and cost.

WASTE PREVENTED: Rework, overprocessing, delays.

EXPLANATION: As explained under subenablers 4.2.4–4.2.9, engineers tend to be proud professionals who like to think they "know what they are doing." Frequently, they execute their assigned tasks according to written specifications—only to discover that the next person in the value chain (the Receiver, a.k.a. internal customer) rejected the task output because it did not exactly conform to the receiver's needs or expectations. In our complex technological world, few tasks are so routine that written specifications are sufficient and no followup verbal clarification and coordination are needed. When a task output is rejected, it must be redone, causing the program to suffer the waste of rework, consuming resources and delaying progress. This waste is notorious in so-called *stovepipe* organizations, in which engineers receive written task specifications from their managers and send output back up to managers for approval and passing on. Frequently, in such organizations, an engineer executing a task (the Giver) does not even know who his or her Receiver is (internal customer other than his or her own manager).

In order to avoid this waste, for every non-routine task, the following is promoted by the subenabler: (1) learn who the internal customer (Receiver) is for your task (it is rarely your manager who should rather serve as an enabling party and a *traffic cop*); (2) coordinate task nuances (scope, modalities, output format, etc.) with the Receiver customer before work on the task begins, reach a clear consensus and minimize bureaucracy; and (3) stay connected to the Receiver to resolve any doubts or questions that may appear during task execution. This will promote execution *right the first time* and avoid rework waste and delays.

In order for this transaction to work, both data deliverer and receiver must understand one another's mutual needs and expectations, basing their expectations on professionalism, teamwork, trust, honesty, openness, and respect. In those rare cases of conflict, issues must be resolved by maximizing value to an end customer.

SUGGESTED IMPLEMENTATION:
- An instruction from a top manager explaining this subenabler and demanding compliance.
- General training in Lean culture.
- Mentoring and good examples.

LAGGING FACTORS:
- Silo organizations, disjointed departments and divisions.
- Work assignments and results flowing vertically (manager–engineer–manager) rather than horizontally, along a value chain, from engineer directly to another engineer.
- Lack of trust, openness, honesty, teamwork, respect.
- Lack of Lean training.

SUGGESTED READING:
- B.W. Oppenheim, Lean product development flow. *Journal of SE*, 2004, **7**, 4, pp. 352–376.
- J.M. Morgan and J.K. Liker, *The Toyota Product Development System*. New York: Productivity Press, 2006.
- A.C. Haggerty and E.M. Murman, *Evidence of Lean Engineering in Aircraft Programs*, 25th *International Congress of the Aeronautical Sciences,* Hamburg, Germany, September 2006, pp. 1–11.

LEAN PRINCIPLE 3: Flow

ENABLER: **7. Make Program Progress Visible to All.**

SUBENABLER(S) AND USE RANKINGS:
1. **Make work progress visible and easy to understand to all, including external customers. (U = 0.36)**
2. **Utilize Visual controls in public spaces for best visibility (avoid computer screens). (U = 0.08)**
3. **Develop a system making imperfections and delays visible to all. (U = 0.16)**
4. **Use traffic light system (green, yellow, red) to report task status visually (good, warning, critical) and make certain problems are not concealed. (U = 0.80)**

VALUE PROMOTED: Adherence to schedule, budgets, and requirements; visibility of progress; visibility of imperfections; motivation for immediate corrective actions.

WASTE PREVENTED: Directly rework and delays, indirectly all other waste types.

EXPLANATION: Experience from Lean companies indicates that visibility of work status and imperfections is conducive to better work flow. "Traditional business processes have the capacity to hide vast inefficiencies without anyone noticing—people just assume that a typical process takes days or weeks to complete. They don't realize that a lean process might accomplish the same thing in a matter of hours or even minutes." [Liker, 2004, 88]. The present four subenablers (3.7.1–3.7.4) promote wide use of the so-called Visual Controls in the PD environment to make all relevant information visible to all, displayed on marker boards in clearly visible spaces. "There are many excellent examples of Visual Controls in everyday life, such as traffic signals and signage. Because it is a matter of life and death, traffic signals tend to be well-designed visual controls. Good traffic signs don't require you to study them: their meaning is immediately clear." [Liker, 2004, p. 152].

"The Toyota Way recognizes that visual management complements humans because we are visually, tactilely, and audibly oriented. And the best visual indicators are right at the work station, where they jab out at you and clearly indicate by sound, sight, feel the standard and any deviation from the standard. Well-developed visual control system increases productivity, reduces defects and mistakes, helps meet deadlines, facilitates communication, improves safety, lowers costs, and generally gives the [employees] more control over their environment." [Liker, 2004, p. 158].

The following are examples of Visual Controls used in PD environment:

- A well-visible marker board hanging just outside each engineer's cubicle, divided into rows and columns; rows listing all task at hand, and columns showing task status (green = task proceeding fine; yellow = minor trouble but the person expects to meet the deadline; red = major issue that may affect the deadline or quality—and a signal to the supervisor: "I may need help"), deadline, corrective actions undertaken, and brief comments. These boards should be updated as soon as the status of any task changes. It takes a few seconds to update them.

- Larger marker boards hanging on a visible wall showing the information important to an entire department (e.g., schedule monitoring, location of employees, perhaps quality trend data, meeting announcements, status of employee training, notes)

- Several large marker boards used in program (project) *war rooms*, showing detailed plans, task responsibilities, milestones, schedules, deadlines, manpower, status of risks, status of supplies, and any other relevant information. Liker [2004, p. 156] describes the visual controls used at Toyota in the design *war room*: "The chief engineer of a vehicle development project resides in the [*war room*], along with heads of major engineering groups working on the project. It is a very large conference *war room* (emphasis of the present author) in which many visual management tools are displayed and maintained by the responsible representatives of the various functional specialties. These tools include the status of each area (and each key supplier) compared with the schedule, design graphics, competitor tear-down results, quality information, manpower charts, financial status, and other important performance indicators. These tools can be reviewed by any of the team members. Any deviation from schedule or performance targets is immediately visible. . . ."

- Green, yellow, and red work status marks (flags, marker signs, magnetized dots, or lights) make imperfections and issues visible to all in real time. An issue that is flagged early tends to be easier to fix. Concealed issues tend to grow to crisis proportions, torpedoing program progress, and requiring heroic actions to resolve.

- Visual controls are highly motivational to best work and to continuous improvement.

- They obviate the need for a manager to ask intimidating questions. Instead the manager seeing the red signal comes to offer help. This alone is conducive to great work morale.

- These controls visually communicate deviations of work status from the expected norm (assigned schedule, quality, etc.)

Some companies implement Visual Controls as computer screens. This is not recommended because it obviates the intended purpose: Individuals see those screens only when looking at them, and most of the time these screens remain invisible concealing the critical information. The goal of Visual Controls is to make all important information visible to all, all the time.

SUGGESTED IMPLEMENTATION: These subenablers are easy to implement. The first step is a short training to explain the benefits of Visual Controls in PD environment to employees and gain their trust. Immediately after training, engineers and managers should brainstorm together in small teams what information to display, where, how, and how often to update it. Managers should resist the strong pull toward computer screens and should insist on posting the information on room and cubicle walls. The company should purchase all necessary materials (marker boards, markers, and color marks). After initial implementation, continuous improvement and 5Ss should be used to keep perfecting the system. The system will be successful when employees are not afraid to display the truth and when managers demonstrate their willingness to help employees when needed.

LAGGING FACTORS:
- Authoritarian management style.
- The culture of secrecy.
- Lack of honesty, openness, trust, respect.

SUGGESTED READING:
- J.K. Liker, *The Toyota Way: 14 Management Principles from the World's Greatest Manufacturer*. McGraw-Hill Professional, 2004.
- E.M. Murman, *Lean Aerospace Engineering*, Littlewood Lecture AIAA-2008-4, January 2008.
- J.M. Morgan and J.K. Liker, *The Toyota Product Development System*. New York: Productivity Press, 2006.
- B.W. Oppenheim, Lean product development flow. *Journal of SE*, 2004, **7**, 4, pp. 352–376.
- A.C. Haggerty and E.M. Murman, *Evidence of Lean Engineering in Aircraft Programs*, 25th *International Congress of the Aeronautical Sciences*, Hamburg, Germany, September 2006.

LEAN PRINCIPLE 3: Flow

ENABLER: 8. Use Lean Tools.

SUBENABLER(S) AND USE RANKINGS:
1. **Use Lean tools to promote the flow of information and minimize handoffs: small batch size of information, small *takt times*, wide communication bandwidth, standardization, work cells, training.** (U = −0.12)
2. **Use minimum number of tools and make common wherever possible.** (U = −0.04)
3. **Minimize the number of the software revision updates and centrally control the update releases to prevent information churning.** (U = 0.36)
4. **Adapt the technology to fit the people and process.** (U = 0.17)
5. **Avoid excessively complex *monument* tools.** (U = −0.04)

VALUE PROMOTED: Shorter schedule, lower cost, and better work flow in programs.

WASTE PREVENTED: Primarily delays and overproduction, but all other categories of waste are also affected.

EXPLANATION: Tool technology should fit people and processes, and not vice versa. Computer tools should aid employees and organizations in improving productivity of PD effort: speeding up the flow of quality work, bringing programs within cost, and minimizing waste. Lean flow of PD work requires minimum handoffs, small batches (packets) of information, short *takt periods*, standardization, teamwork of co-located engineers (i.e., IPTs), and good training. Good software tools should promote these characteristics. Instead, many software tools seem to do the opposite: They are exceedingly complex and difficult to learn and use, expensive to implement, and error prone. This causes rework, promotes large inventories of data and poor interpersonal communications, and the tools actually slow program progress.

Since each new tool requires implementation and learning costs, the number of tools in use should be kept low.

Implementation of each new release of a complex software carries with it costs, delays, learning curves, and possibly changes in standards and processes. If one and the same tool is used across several organizations and suppliers, the new releases should be coordinated across all stakeholders. If a tool maker introduces frequent releases, program employees may waste a significant fraction of program schedule and budget just learning and adapting to new software at the expense of value adding time. For this reason, the number of update releases should be kept at minimum. It is much more important that all program stakeholders use the same release version than for some to have the latest version that is not compatible for the rest.

Ideally, all released data in a program should be kept in a single electronic database with efficient release system, using user friendly configuration management.

Lean manufacturing uses the term *monument* to sarcastically describe a large, expensive and complex machine that is faster than what a flow needs—it is a waste. Such machines consume excessive resources on learning, maintenance, energy, repairs, and so on. Similarly, in PD, a *monument* tool denotes *the latest, greatest, and gold-plated* software tool that has too many options, is too complex, and is too expensive to buy, learn, and use. If Excel can do the job in five minutes, why use hugely complex *monument* software that takes one day to complete the same task?

The following is an outline of effective steps adopted under Principle 11 (Avoid excessively complex *monument* tools) in Morgan and Liker [2006, pp. 242–243]:

1. "Technologies must be seamlessly integrated.
2. Technologies should support the process, not drive it.
3. Technologies should enhance people, not replace them.
4. Specific solution oriented: not a silver bullet.
5. Right size—not king sized."

SUGGESTED IMPLEMENTATION: Selection of tools requires expert knowledge of the PD domain and of the SE processes. Experts should perform careful studies of available tools on the market and match them to the needs of enterprise stakeholders, including major suppliers, and customer preferences. Experts should also understand Lean to be able to make optimum choices.

LAGGING FACTORS:
- Mixed heritage of organizations, each bringing its separate and incompatible tool suite into the merger*.
- Government push for *the latest, greatest, and gold-plated*.
- Rushed selection of tools.
- Selection of tools that "do everything."
- Lack of understanding of stakeholder needs and work flow needs.

---

*After the mega-mergers of defense companies in the mid 1990s, a Vice President of one joked to the present author that the merged company had already spent close to $1 billion on software tool standardization, and was not done yet.

SUGGESTED READING:
- J.K. Liker, *The Toyota Way: 14 Management Principles from the World's Greatest Manufacturer*. McGraw-Hill Professional, 2004.
- J.M. Morgan and J.K. Liker, *The Toyota Product Development System*. New York: Productivity Press, 2006.
- D. Lempia, *Using Lean Principles and MBE in Design and Development of Avionics Equipment at Rockwell Collins*, 26th *International Congress of the Aeronautical Sciences,* Anchorage, AK, September 16, 2008, paper 6.7.3.

7.2.4   Lean Principle 4: Pull

"An unreleased [...] study has found that an alarming percentage of PD process outputs are not needed by downstream processes, for program knowledge capture, for meeting regulations, contractual requirements, or quality standards, or for any other purpose. They are waste." [McManus, 2004].

Summary   Two enablers are listed under this Principle, as follows. The numbers in parentheses list the number of subenablers under each enabler.

Enabler 4.1 Tailor for a Given Program According to the INCOSE Handbook Process.

The first enabler listed under each principle emphasizes the same fundamental point: to add the wisdom of Lean Thinking to the traditional Systems Engineering process rather than replace the process. So, again, the present enabler recommends to continue using the INCOSE SE Handbook, or any other equivalent SE handbook or manual, and in addition to implement the enablers that follow.

Enabler 4.2 Pull Tasks and Outputs Based on Need and Reject Others as Waste (9)

The "Pull" Principle is a powerful guard against the waste of unneeded tasks, overprocessed tasks, task rework (not to be confused with legitimately needed and optimized iteration loops), and tasks that are left over from previous programs or company habits. Pull promotes tailoring tasks and pulling them and their outputs based on legitimate need, rejecting others as waste ("legitimate" is interpreted in the context of Lean SE value definition: "as needed for flawless system success or mission assurance"). Each task in the flow should be needed by some program stakeholder, either internal or external customer. Any task for which the team cannot find a customer should be classified as a waste. Experience from recent complex programs indicates that the number of unneeded tasks is significant. Some tasks may have been ordered in a previous program, and the requester forgot to indicate that this is only a single use of task. A manager may have ordered a special task for his or her own special application, and this manager is no longer employed, yet the task is repeated in subsequent programs. Many unneeded tasks end up in the schedule because an inexperienced person uncritically performs a cut–and–paste operation from a previous program to the current program.

The following tables describe the Principle 4 subenablers individually.

LEAN PRINCIPLE 4: Pull

ENABLER: 2. Pull Tasks and Outputs Based on Need and Reject Others as Waste.

SUBENABLER(S) AND USE RANKINGS:
1. **Let information needs pull the necessary work activities.** (U = −0.04)
2. **Promote the culture in which engineers pull knowledge as they need it and limit the supply of information to only genuine users.** (U = −0.04)
3. **Understand the Value Stream Flow.** (U = −0.32)

VALUE PROMOTED: Shorter and less costly programs.

WASTE PREVENTED: Execution of unneeded tasks.

EXPLANATION: As indicated by low use rankings, recent complex programs are notorious for including tasks that nobody needs, that perhaps some manager ordered some time ago for a particular reason, the reason long forgotten. Many tasks are included by careless cutting-and-pasting from earlier programs. All such tasks constitute pure waste. Therefore, rigorous tailoring of tasks and outputs should be done in the planning phase of every program. The cardinal rules should be:
1. Identify the genuine users for each task output and have them pull information as needed and supply information only to those who need it.
2. If no user can be identified for a given task output, the task is a waste and should be eliminated.

"The speed at which technology delivers information in . . . product development is overwhelming. However, not all information is equal to all people. The lean PD System uses 'pull' to sort through this mass of data to get the right information to the right engineer at the right time." [Morgan and Liker, 2006, p. 96].

See also subenablers 2.2.10 and 4.2.4 describing the RASI and SIPOC planning approaches.

SUGGESTED IMPLEMENTATION: The implementation is quite easy; all it takes is an instruction from the Chief SE issued at the beginning of the planning phase to practice the subenablers listed above (4.2.1–4.2.3), emphasizing the following:
- Do not copy tasks from previous programs to present program carelessly.
- If you cannot find a user for a task output, delete the task.
- The task is legitimate if it supports value creation.

LAGGING FACTORS:
- Rushed ad-hoc planning.
- Lack of understanding of Value-Stream Mapping techniques for PD.
- Careless insertion of tasks from former programs, regardless of actual need.

SUGGESTED READING:
- H.L. McManus, *Product Development Value Stream Mapping Manual*. LAI Release Beta, Massachusetts Institute of Technology, LAI, April 2004.
- J.M. Morgan and J.K. Liker, *The Toyota Product Development System*. New York: Productivity Press, 2006.
- D. Lempia, *Using Lean Principles and MBE in Design and Development of Avionics Equipment at Rockwell Collins*, *26th International Congress of the Aeronautical Sciences,* Anchorage, AK, September 16, 2008, paper 6.7.3.
- D. Secor, *Implementation of Lean Enablers for Systems Engineering at Rockwell Collins*, LAI Knowledge Exchange Event, LAI Plenary, March 23, 2010.

LEAN PRINCIPLE 4: Pull

ENABLER: 2. Pull Tasks and Outputs Based on Need and Reject Others as Waste.

SUBENABLER(S) AND USE RANKINGS:
4. **Train the team to recognize who the internal customer (Receiver) is for every task as well as the supplier (Giver) to each task—use a SIPOC (supplier, inputs, process, outputs, customer) model to better understand the value stream. (U = −0.04)**
5. **Stay connected to the internal customer during the task execution. (U = 0.32)**
6. **Avoid rework by coordinating task requirements with internal customer for every non-routine task. (U = 0.08)**
7. **Promote effective real time direct communication between each Giver and Receiver in the value flow. (U = 0.24)**
8. **Develop Giver-Receiver relationships based on mutual trust and respect. (U = 0.36)**
9. **When pulling work, use customer value to delineate value adding work from waste. (U = 0.00)**

VALUE PROMOTED: Faster and *right the first time* completion of tasks.

WASTE PREVENTED: Rework.

EXPLANATION: As mentioned under subenabler 3.6.6, experienced engineers are proud professionals who like to think that they "know what they are doing." Frequently, they execute their assigned tasks according to written specifications, only to discover that the next person in the value chain (the Receiver, a.k.a. internal customer) rejects the task output because it does not exactly conform to the Receiver's needs or expectations. In our complex technological world, few tasks are so routine that written specifications are sufficient, and no followup verbal clarification and coordination is needed. When a task output is rejected, it must be redone, causing the program to suffer rework. This waste is notorious in so-called stove-piped organizations, in which an engineer receives written task specifications and sends output back up to his or her manager for approval and passing on. Frequently, in such organizations, the engineer executing the task (the Giver) does not even know who his or her Receiver is.
In order to avoid this waste, for every non-routine task, the above subenablers (4.2.4–4.2.9) promote the discipline of every Giver to:
1. Learn who the Receiver (internal customer) is for the task (it is rarely the manager—who should rather serve as an enabling party and a *traffic cop*).
2. Coordinate the task (scope, modalities, output format, etc.) with that internal customer before the work begins until clear consensus is achieved, with minimum bureaucracy.

3. Stay connected to the Receiver to resolve any doubts or questions that may appear during task execution. This will promote execution *right the first time* and avoid rework waste.

In order to avoid conflicts, both Giver and Receiver must base their interactions on professionalism, trust, and respect.

The Work Breakdown Structure practiced in many program planning efforts is usually not detailed enough. A better system is to apply the SIPOC model, defining for each task the Giver, inputs, process, outputs, the internal customer (a.k.a. the Receiver) to better understand the value stream.

In all Giver–Receiver transactions, there is no substitute for mutual trust and respect. Both are needed to accept Receiver's wishes with confidence that they are indeed legitimate, needed for overall value, and not a whim of the person. In case of rare conflict, the issue must be resolved by maximizing value to the end customer.

SUGGESTED IMPLEMENTATION: These subenablers are easy to implement by a short training session explained under subenabler 3.6.6, or a written instruction from the top program manager, disseminated to stakeholders during program preparation phase. The instructions should be followed by mentoring and good examples.

LAGGING FACTORS:
- Traditional authoritarian culture where work is assigned by a manager, and the task output goes back to the manager, with the Giver not knowing who is the Receiver, or why the Receiver needs the information.
- Ad-hoc planning.
- Planning with too low a resolution, e.g., using only the Work Breakdown Structure.
- *Stovepipe* organizations.

SUGGESTED READING:
- B.W. Oppenheim, Lean product development flow, *Journal of SE*, 2004, **7**, 4, pp. 352–376.
- J.M. Morgan and J.K. Liker, *The Toyota Product Development System*. New York: Productivity Press, 2006.
- D. Lempia, *Using Lean Principles and MBE in Design and Development of Avionics Equipment at Rockwell Collins*, 26th *International Congress of the Aeronautical Sciences,* Anchorage, AK, September 16, 2008, paper 6.7.3.

7.2.5   Lean Principle 5: Perfection

Summary   The fifth or "Perfection" Principle strives for excellence and continuous improvement of SE processes and related Enterprise Management. The listed enablers promote:

- Making all imperfections visible to all, which is motivating for immediate improvement.
- Comprehensive capture and use of lessons learned from past programs.
- Driving out waste through design standardization, process standardization, and skill-set standardization.
- Employing all three mutually complementary Continuous Improvement (CI) methods: bottom-up suggestion system; quick-reaction, small *Kaizen* teams for fixing local problems; and a formal Six Sigma approach for large complex problems.
- Excellent communication, coordination, and collaboration to enable CI.
- Elevation of the role of Chief SE to lead and integrate programs from start to finish.

Seven enablers are listed under this Principle. The numbers in parentheses list the number of subenablers.

Enabler 5.1 Pursue Continuous Improvement According to the INCOSE Handbook Process.

The first enabler listed under each principle emphasizes the same fundamental point: to add the wisdom of Lean Thinking to the traditional Systems Engineering process rather than replace the process. So, again, the present enabler recommends the continuation in use of the INCOSE SE Handbook, or any other equivalent SE Handbook. It also recommends implementing the enablers that follow.

Enabler 5.2 Strive for Excellence of SE Processes. (8)

Continuous improvement of process is often confused with continuous improvement of process content. Process improvements must be pursed continuously for all important processes if an enterprise is to remain competitive. New knowledge, methods, tools, procedures, checklists, and lessons learned change all the time and they must be incorporated into the SE process as soon as possible. In contrast, it is not practical to continuously refine process content, which could break program budget and schedule. In Lean, it is important to optimize value but not to sub optimize its elements. Subenablers 5.2.6–5.2.7 recommend how to set limits on the refinements.

Enabler 5.3 Use Lessons Learned from Past Programs for Future Programs. (5)

The five subenablers listed here promote making each program better then the last, and creating effective mechanisms to capture and apply experience-generated learning and checklists.

Enabler 5.4 Develop Perfect Communication, Coordination and Collaboration Policy Across People and Processes. (8)

Effective and efficient communications represent one of the most critical aspects of good programs. Good communications and coordination are needed among people within and between processes.

Many programs practice only ineffective ad-hoc communications. The eight subenablers listed here deal with several aspects of excellent communications: communication policy formulation, planning and training, directives for email and artifact management, use of A3 forms, selection of individuals on the basis of good communication skills, and creation of databases.

Enabler 5.5 For Every Program Use a Chief Engineer Role to Lead and Integrate the Program from Start to Finish. (5)

The five subenablers promote a single qualified person to be fully responsible, accountable, and with authority for entire technical program success, following best U.S. historical examples and many international examples.

Enabler 5.6 Drive Out Waste Through Design Standardization, Process Standardization, and Skill-set Standardization. (3)

The three subenablers address the best standardization practices, including standardization of designs, processes, and engineering skill sets.

Enabler 5.7 Promote All Three Complementary Continuous Improvement Methods to Draw Best Energy and Creativity From All Employees. (3)

Organizations must have effective means of preventing and addressing all problems, frustrations, wastes, complaints, and so on. For competitive reasons and for the internal health of an organization, continuous improvement (CI) must be conducted for all activities (subject to reasonable limits on overprocessing waste) in the spirit that everything can always be done in a better way, subject to wise prioritization. The three subenablers promote three mutually complementary CI methods: a *bottom-up* employee suggestion system, a small quick-reaction CI team called *Kaizen*, and a powerful solution to large problems, called Six Sigma.

The following tables describe the Principle 5 subenablers individually.

LEAN PRINCIPLE 5: Perfection

ENABLER: 2. Strive for Excellence of SE Processes.

SUBENABLER(S) AND USE RANKINGS:
1. Do not ignore the basics of Quality (U = 0.84):

    a. **Build in robust quality at each step of the process, and resolve and do not pass along problems (U = 0.17)**

    b. **Strive for perfection in each process step without introducing waste. (U = −0.16)**

    c. **Do not rely on final inspection; error proof wherever possible. (U = 0.08)**

    d. **If final inspection is required by contract, perfect upstream processes pursuing 100% inspection pass rate. (U = 0.28)**

    e. **Move final inspectors upstream to take the role of quality mentors. (U = 0.08)**

    f. **Apply basic PDCA method (plan, do, check, act) to problem solving. (U = 0.48)**

    g. **Adopt and promote a culture of stopping and permanently fixing a problem as soon as it becomes apparent. (U = −0.08)**

2. Promote excellence under "normal" circumstances instead of hero-behavior in "crisis" situations. (U = −0.40)

VALUE PROMOTED: Predictable and uninterrupted flow of work, the least costly way to execute programs.

WASTE PREVENTED: Rework, waiting, loss of competition, market failure.

EXPLANATION: These subenablers (5.2.1a-g and 5.2.2) collectively repeat the critically important lessons from the Total Quality Management period that have been fully embraced by Six Sigma and Lean. The lessons include global competitive pressures to provide excellent quality inexpensively. The main lesson of TQM is that processes that are unpredictable in terms of quality, time, effort, and cost can torpedo program success. When Womack et al. [1990] conducted their automobile production research in Europe, it didn't take them long to find the underlying problem afflicting that industry: "A widespread conviction among managers and workers that they were craftsmen. At the end of the assembly line was an enormous rework and recertification area where armies of technicians in white laboratory jackets labored to bring the finished vehicles up to the company's fabled quality standard. We found that a third of the total effort involved in assembly occurred in this area. In other words, the German plant was expending more effort to fix the problem it had just created than the Japanese plant required to make a nearly perfect car the first time" [Womack et al., 2006, pp. 90–91]. The same can be said about many PD efforts, including Systems Engineering.

The subenablers listed in this table promote the following classical approaches to quality:

a. Design each step of each process to be consistently predictable and robust (creating repeatable good quality regardless of normal operational conditions). And, if an imperfection is found, empower employees to stop the process, find the root cause, and fix it right away. Never pass problems along. If everyone in the value chain would follow this subenabler (5.2.1a), the final quality inspectors might not even be needed.

b. Do not rely on final inspection because it is never reliable (some defects always manage to escape notice and get passed on to the customer) and is always the most costly of all quality approaches. This subenabler (5.2.1b) was first suggested by Deming [1982] is his third Point: "Cease dependence on mass inspection: Eliminate the need for mass inspection as the way of life to achieve quality by building quality into the product in the first place."

c. Keep refining each process until 100% pass rate is assured (that is, *right the first time*), and the process is predictable and robust.

d. If final inspection is required by contract, perfect the upstream processes until a 100% quality pass rate is achieved.

e. Apply the classical tools of Continuous Improvement (CI), of which the *Plan-Do-Check-Act (PDCA)* cycle is a proverbial sequence.

f. When quality inspectors catch imperfections, they should mentor the upstream processes how to eliminate these imperfections.

g. Adopt and promote a culture of stopping and permanently fixing a problem as soon as it becomes apparent. Empower employees at every level to stop a flow if any imperfections are detected. Then perform a root cause analysis and permanently fix the problem. Then adjust the process, checklists, training, and procedures to ensure defects will not be repeated.

Experience indicates that it is always cheaper and less frustrating to pursue excellence as a matter of routine, which obviates the need for heroic actions later, trying to fix a problem that was left unattended when it was small and easy to fix. The following quote illustrates the quality challenge: "If pure water flows from the upper stream, there is no need to purify it farther downstream" (attributed to Hiroto Kagami, Senior Manager of Quality Assurance at Cannon).

No company can afford to invest unlimited funds and time into continuous improvement alone. Subenablers 5.2.6–5.2.7 address this concern and recommend how to set the limits on improvements.

Toyota has been the standard bearer of high quality in all areas of Enterprise: manufacturing, PD, SE, supply chain, and support functions.

SUGGESTED IMPLEMENTATION: The transformation from a traditional company where quality is treated as a luxury; processes are executed as a craft, lacking predictability and robustness; where rework is notorious and massive final inspection is needed to Lean is long and difficult, but achievable, as demonstrated by numerous industrial examples. This challenge is beyond the scope of the present text and the readers are referred to a large library of textbooks on quality improvements.

LAGGING FACTORS:
- Never enough time to fix problems.
- Culture of "get it out the door by any means and we will fix it later."
- Allowing the same problem to occur more than once.
- Lack of training.
- Lack of competitive pressures.

SUGGESTED READING:
- J.P. Womack and D.T. Jones, *Lean Thinking*, Simon & Shuster, 1996.
- J.K. Liker, *The Toyota Way: 14 Management Principles from the World's Greatest Manufacturer*. McGraw-Hill Professional, 2004.
- W.E. Deming, *Out of the Crisis*. Massachusetts Institute of Technology, Center for Advanced Engineering Study, 1982.
- J.M. Morgan and J.K. Liker, *The Toyota Product Development System*. New York: Productivity Press, 2006.
- R. Leopold, *The Iridium Story: An Engineer's Eclectic Journey*, Minta Martin Lecture, MIT Department of Aeronautics and Astronautics, April 23, 2004.
- C. Baden-Fuller and J.M. Stopford, *Rejuvenating the Mature Business: The Competitive Challenge*, Routledge, 1992.

LEAN PRINCIPLE 5: Perfection

ENABLER: 2. Strive for Excellence of SE Processes.

SUBENABLER(S) AND USE RANKINGS:
3. **Use and communicate failures as opportunities for learning emphasizing process and not people problems.** (U = 0.04)
4. **Treat any imperfection as an opportunity for immediate improvement and lesson to be learned and practice frequent reviews of lessons learned.** (U = −0.20)

VALUE PROMOTED: Continuous improvement of the SE process and of company competitiveness.

WASTE PREVENTED: All categories of waste.

EXPLANATION: The Perfection Principle implies profound knowledge on how to deal with imperfections and failures. Every imperfection should be made visible to all in real time, so that improvement can be undertaken immediately. Employees who are not afraid of being penalized or fired feel motivated to identify all imperfections immediately. An imperfection should be treated as an opportunity to improve the system. If concealed or ignored, imperfections tend to grow to crisis proportions, later requiring heroic and expensive measures to save the program. Once an imperfection is discovered, an immediate root cause analysis and improvement should be undertaken. Next, a serious effort should be made to treat the imperfection as a lesson to be learned and to disseminate it to the team. Also, the applicable checklists, procedures, and training should be updated so that the particular imperfection will never be allowed to occur again.
All failures and issues manifest themselves personally (it is a person who brings the news to the attention of stakeholders), yet they almost always represent faulty processes. Good managers understand this and do not blame the messengers or process operators, but rather mentor in problem elimination, making sure it never occurs again. The working hypothesis should be that a process has been designed well, including tools and training, and if the process failed, it is the fault of the process designer and trainer (i.e., the manager responsible) rather than the operator. E. Deming [1982] never tired of pointing out that "85% of all problems are caused by management and only 15% by line employees."
Deming taught that every normal employee is eager to work well. The exceptions are the few percents of pathological cases (drugs, alcohol, absences, etc.). The system should be designed to motivate the best work from the vast majority of employees rather than police the small minority. This promotes a right culture based on teamwork, openness, honesty, and respect. In this culture employees are motivated to look for and identify imperfections in real time for immediate improvements. Such an organization becomes a perpetual improvement system becoming better all the time.

"In striking contrast to the mass-production plant, where stopping the line was the responsibility of the senior line manager, Taiichi Ohno (Toyota's chief production engineer) placed a cord above every worker station and instructed workers to stop the whole assembly line immediately if a problem emerged that they couldn't fix. Then the [...] team would come over to work on the problem." [Womack et al., 2007, p. 56]. The same line-stopping culture must be routinely practiced in Lean SE.

SUGGESTED IMPLEMENTATION: Promote a work culture based on trust, openness, honesty, and teamwork. Eliminate those elements that antagonize employees and perpetuate fear of superiors and communication. Train and mentor employees in the benefits of the new work culture and continuous improvement. Motivate employees to make imperfections and failures visible and to identify them in real time. Seek root causes and permanent fixes for all problems. Develop a system focused on motivating the vast majority of employees for best work, rather than policing the small minority of non-performers. Encourage engineers and other employees to bring their suggestions for improvement to managers and do your best to act on these suggestions. Promote updating checklists, standards, and training immediately with lessons learned.

LAGGING FACTORS:
- Fear and lack of trust, openness, honesty, teamwork.
- Rushing through corrective actions without enough time to learn the lessons.

SUGGESTED READING:
- J.P. Womack and D.T. Jones, *Lean Thinking*, Simon & Shuster, 1996.
- J.M. Morgan and J.K. Liker, *The Toyota Product Development System*. New York: Productivity Press, 2006.
- W.E. Deming, *Out of the Crisis*. Massachusetts Institute of Technology, Center for Advanced Engineering Study, 1982.
- D. Lempia, *Using Lean Principles and MBE in Design and Development of Avionics Equipment at Rockwell Collins*, 26th *International Congress of the Aeronautical Sciences,* Anchorage, AK, September 16, 2008, paper 6.7.3.
- R. Leopold, *The Iridium Story: An Engineer's Eclectic Journey*, Minta Martin Lecture, MIT Department of Aeronautics and Astronautics, April 23, 2004.

LEAN PRINCIPLE 5: Perfection

ENABLER: 2. Strive for Excellence of SE Processes.

SUBENABLER(S) AND USE RANKINGS:
5. Maintain a consistent disciplined approach to engineering. (U = 0.52)

VALUE PROMOTED: Robust flow, *right the first time*, predictable schedule.

WASTE PREVENTED: Rework, overproduction, waiting.

EXPLANATION: Even though engineering is one of the oldest professions, much of the practice is still executed as a craft: Each task is treated as a unique challenge to be planned from scratch and executed according to the particular experience and knowledge of an engineer in charge. Engineers often confuse process with process content. While the latter is indeed always unique, the former should be standardized and continually optimized. This is true for most analysis and test processes, and even for design processes. This yields a disciplined approach to engineering, establishing and repeatedly following established best practices, leading to predictable and robust and optimized flow. The concept that most of engineering can and should be organized into processes and checklists capturing best engineering practices is often poorly understood, ignored, and sometimes even resisted. This subenabler (5.2.5) states that for standard tasks (such as a structural analysis of a part), a process and/or a checklist should be created and followed. In order for this to work, checklists and processes should objectively represent best engineering practices. They should be frequently updated based on the advancements, lessons learned, and continuous improvement opportunities. They must be user friendly, and stakeholders should be well trained in all procedures and checklists. The number of checklists and procedures should be small enough to improve the quality of work, and not overwhelm engineers with bureaucracy.
Indeed, some portion of the engineering effort cannot be fitted into standard processes and checklists. However, even in complex high-tech programs, the effort that can and should be standardized is significant and worth implementing.
One reason why engineers often resist standards is psychological: Engineers see themselves as creative problem solvers and dislike having to follow in somebody's footsteps. It is important, therefore, to convince engineers of the tangible benefits that come from disciplined approach: quicker completion of standard tasks and more time for creative ideas.

SUGGESTED IMPLEMENTATION:
- Training for new hires about the benefits and the use of the best engineering checklists and procedures.
- The checklists and procedures must be prepared well and user friendly or they will be ineffective and ignored. See also subenablers 5.6.1–5.6.3.

LAGGING FACTORS:

- The culture of treating all engineering work as unique.
- Lack of disciplined approach to engineering.
- Confusing process with process content.
- Culture of endless refinement of product, beyond what constitutes value (as opposed to the process, which should be continually improved; see subenablers 5.2.6–5.2.7).

SUGGESTED READING:

- B.W. Oppenheim, Lean product development flow, *Journal of SE*, 2004, **7**, 4, pp. 352–376.
- J.M. Morgan and J.K. Liker, *The Toyota Product Development System*. New York: Productivity Press, 2006, pp. 27–37, 339.

LEAN PRINCIPLE 5: Perfection

ENABLER: 2. Strive for Excellence of SE Processes.

SUBENABLER(S) AND USE RANKINGS:

6. **Promote the idea that the system should incorporate continuous improvement in the organizational culture, but also ...(continued in 5.2.7) (U = 0.42)**

7. **... balance the need for excellence with avoidance of overproduction waste (pursue refinement to the point of assuring value and *right the first time*, and prevent overprocessing waste). (U = 0.25)**

VALUE PROMOTED: Minimum resources spent on creating value.

WASTE PREVENTED: Overprocessing.

EXPLANATION: A frequent question asked during design/analysis/test is, "How much refinement is good enough?" The tradeoff is between added excellence, on the one hand, and program schedule and budget on the other. Typically, engineers seek perfection while Systems Engineers responsible for budget and schedule seek a "good-enough" level. These two subenablers suggest a positive compromise.

Pursue the perfection of a repeated process up to the point of achieving a process that consistently and predictably executes *right the first time* and is competitive. Pursue the perfection or refinement of the system or mission until the needs of the value proposition have been met. The waste of overprocessing begins immediately beyond these levels of excellence.

SUGGESTED IMPLEMENTATION: These subenablers are easy to implement in the two following steps:
- Explain the above to all engineers.
- Mentor engineers in these subenablers until they are well practiced.
- Empower Systems Engineers to demand adherence to these subenablers from all other engineers.

LAGGING FACTORS:
- Uncontrolled refinements by engineers, in excess of value proposition, disregarding overprocessing waste.

SUGGESTED READING:
- P.B. Crosby, *Let's Talk Quality: 96 Questions You Always Wanted to Ask Phil Crosby*. McGraw-Hill Publishing Company, New York, 1989.
- D. Secor, *Implementation of Lean Enablers for Systems Engineering at Rockwell Collins*, LAI Knowledge Exchange Event, LAI Plenary, March 23, 2010.
- D. Lempia, *Using Lean Principles and MBE in Design and Development of Avionics Equipment at Rockwell Collins*, 26th *International Congress of the Aeronautical Sciences,* Anchorage, AK, September 16, 2008, paper 6.7.3.
- J.M. Morgan and J.K. Liker, *The Toyota Product Development System*. New York: Productivity Press, 2006.

LEAN PRINCIPLE 5: Perfection

ENABLER: 2. Strive for Excellence of SE Processes.

SUBENABLER(S) AND USE RANKINGS:
8. Use a balanced matrix/project organizational approach avoiding extremes: territorial functional organizations with isolated technical specialists, and all-powerful IPTs separated from functional expertise and standardization. (U = 0.21).

VALUE PROMOTED: Optimal organization of the matrix, better programs.

WASTE PREVENTED: All categories of waste.

EXPLANATION: Programs must be properly balanced between functions (departments) and Integrated Product Teams (IPTs). Functions are better connected to technology changes and progress, but are inherently less connected to projects and customers. IPTs are intended to serve a particular program and its customer better, but tend to be disconnected from cutting-edge technology changes. Both are needed, but the relative strengths of the two should be optimized for a particular program. Thomas Allen [2002] proposes an optimization based on the following variables:
- Rate of change of technology (if the rate is fast, give preference to stronger functions and weaker IPTs).

- Program duration (if long, strong IPTs serve programs better).

- Subsystem interdependence (if high, strong IPTs work better).

- Rate of change of market volatility (if high, strong IPTs work better).

Competition for resources in matrix organizations is natural and must be dealt with effectively. Some corporations establish a position of a neutral arbiter that negotiates and prioritizes limited human resources among competing programs, based on the overall needs of the entire program portfolio.

SUGGESTED IMPLEMENTATION: Most modern project-oriented companies practice a matrix organization, but not many optimize the relative strengths of functions and IPTs well. It is suggested that a small group of Systems Engineering managers adopt the Allen optimization method. A devoted *Kaizen* team comprised of Lead Systems Engineers from different programs should be able to formulate this policy.

LAGGING FACTORS:
- Strong functional managers who are not willing to cooperate well with IPTs.
- Strong IPT managers disconnected from functions.
- Ad-hoc allocation of resources.

SUGGESTED READING:

- J.M. Morgan and J.K. Liker, *The Toyota Product Development System*. New York: Productivity Press, 2006.
- T. Allen, *Optimization of Organizational Matrix*, MIT LAI, Plenary Conference, Los Angeles, 2002.
- A.C. Haggerty, *The F/A-18E/F Super Hornet as a Case Study in "Value Based" Systems Engineering*, INCOSE 2004, Toulouse, France, June 20–24, 2004.

LEAN PRINCIPLE 5: Perfection

ENABLER: 3. Use Lessons Learned from Past Programs for Future Programs.

SUBENABLER(S) AND USE RANKINGS:
1. **Maximize opportunities to make each next program better then the last. (U = 0.13)**
2. **Create mechanisms to capture, communicate, and apply experience-generated learning and checklists. (U = 0.17)**

VALUE PROMOTED: *Right the first time* execution, robust flow, savings in cost and schedule, compliance with standards.

WASTE PREVENTED: All categories of waste.

EXPLANATION: Checklists represent a powerful tool for capturing lessons learned and improving programs, as pointed by [Sobek et al., 1998, pp. 36–49]: "Engineering checklists contain detailed information concerning any number of aspects, including functionality, manufacturability, government regulations, and reliability. Engineers use the checklists to guide the design throughout the development process. The checklists are particularly important for intensive design reviews that every vehicle program undergoes. What keeps these extremely large meetings from becoming chaotic is that all engineers come with a list of all the items they need to verify from their perspectives. If the design conforms to the checklist, the part is highly likely to meet a certain level of functionality, manufacturability, quality, and reliability. If it does not, discrepancies between the checklist and the design become the focal points of discussion among the divisions."

Morgan and Liker [2006, p. 205] point out the importance of the experience-generated learning and tacit knowledge. Paraphrasing the authors, most companies focus on *explicit* 'know what' knowledge–such as mathematical equations and dates—which is usually stored as facts, symbols and easily conveyed without loss of integrity. In addition to the explicit, there is *tacit* **'know how'** (emphasis by present author) knowledge also known as 'sticky' knowledge. Tacit knowledge involves the unspoken transfer of knowledge through long and deep relationships between teacher and student. ... "One of the main reasons that companies fail at imitating lean systems is that they mistakenly copy only the 'explicit' knowledge of lean tools and techniques. By and large, these companies attempt to implement lean without understanding the need to tap into the tacit knowledge of lean culture, the **know-how** knowledge that enables an organization to learn organically, adapt, and grow. They fail to grasp that in highly technical environments, such as product development, tacit knowledge is the true source of competitive advantage."

SUGGESTED IMPLEMENTATION:
- Capture immediately all lessons learned into effective checklists, procedures, and databases.
- Initiate, mentor, promote, and enforce the use of checklists.
- Make sure checklists and procedures are user friendly, or they will be ignored.
- Perform a short (e.g., 5 min) training about the changes describing what is being changed, why, and how.
- Make sure bureaucracy is not smuggled under the cover of checklists and procedures.
- Make certain checklists and procedures can be updated easily without excessive bureaucracy.

LAGGING FACTORS:
- Never enough time or budget to implement improvements.
- Checklists not used.

SUGGESTED READING:
- D.K. Sobek II et al., Another look at Toyota's integrated product development, *Harvard Business Review*, July–August 1998, **76**, 4, pp. 36–49.
- J.M. Morgan and J.K. Liker, *The Toyota Product Development System*. New York: Productivity Press, 2006.
- A.C. Haggerty, *The F/A-18E/F Super Hornet as aCcase Study in "Value Based" Systems Engineering*, INCOSE 2004, Toulouse, France, June 20–24, 2004.

LEAN PRINCIPLE 5: Perfection

ENABLER: 3. Use Lessons Learned from Past Programs for Future Programs.

SUBENABLER(S) AND USE RANKINGS:
3. Insist on workforce training of root cause and appropriate corrective action. (U = 0.04)

VALUE PROMOTED: *Right the first time* execution of processes. Previous mistakes and imperfections not repeated.

WASTE PREVENTED: Rework, delays, and other categories of waste.

EXPLANATION: Rushing a program without any time for implementing improvements tends to invite defect repetitions. In contrast, the habit of finding root cause for every problem and applying immediate corrective action tends to assure that the same mistake will not be repeated. In the long term, the second approach yields more competitive enterprises.

Taiichi Ohno, Toyota's chief production engineer, instituted a system of finding root cause called *five whys*. "Production workers were taught to trace systematically every error back to its ultimate cause (by asking 'why' as each layer of the problem was uncovered), then to devise a fix, so that it would never occur again." [Womack et al., 1996]. Toyota's experience suggests that, on average, *five whys* are needed to help determine the root cause of most technical problems, thus the name. A short training is needed in the *five whys*, but the training alone will not be sufficient. Also needed is the habit of immediately stopping work when an imperfection is found, organizing a small team (or only oneself) to find the root cause, and applying an effective corrective action. For simple local problems the process owner alone is often sufficient. For larger local problems that involve several stakeholders a quick-reaction small *Kaizen* team may work best. For large problems a formal Six Sigma approach based on statistics may be required.

According to Harvard Business Review [HBR, 2008] "Managers should coach, not fix." The result of this manager-worker relationship is "a high degree of sophisticated problem solving at all levels of the organization... any operating system can be improved if enough people at every level are looking and experimenting close enough. ...."

SUGGESTED IMPLEMENTATION:
- Training of employees at all levels in modern quality methods, including the *five whys* and root cause analysis.
- Mentoring, empowerment, and the culture of stopping the line when an imperfection is found.

LAGGING FACTORS:
- Rushing through programs without enough time to learn from imperfections.
- Lack of training in *five whys* and root cause analysis.

SUGGESTED READING:
- *Manufacturing Excellence at Toyota*, Harvard Business School Series, 2008.
- J.M. Morgan and J.K. Liker, *The Toyota Product Development System*. New York: Productivity Press, 2006.
- J.P. Womack and D.T. Jones, *Lean Thinking*, Simon & Schuster, 1996.
- A.C. Haggerty, *The F/A-18E/F Super Hornet as a Case Study in "Value Based" Systems Engineering*, INCOSE 2004, Toulouse, France, June 20–24, 2004.

LEAN PRINCIPLE 5: Perfection

ENABLER: 3. Use Lessons Learned from Past Programs for Future Programs.

SUBENABLER(S) AND USE RANKINGS:
4. Identify best practices through benchmarking and professional literature. (U = 0.26)

VALUE PROMOTED: Improve long-term competitive position of the company.

WASTE PREVENTED: All categories of waste.

EXPLANATION: Competitive pressures in the global economy require all companies to continually improve their methods, tools, processes, products, and people. Benchmarking is an effective way to gain legal knowledge about one's competition. This is carried out through collecting information about the competition using methods such as attending industrial fairs, reading professional literature, performing reverse engineering on competitor's products, networking at professional meetings, and visits with cooperating competitors. Whatever the method used, the goal is the same: to learn what constitutes the state of the art and how to implement and even overtake it. When benchmarking, it is important to remember that the relative positions of competitors are not static. Competitor positions change all the time; therefore, it is essential that the rate of improvement be also benchmarked. Latest inventions and designs are trade secrets beyond benchmarking. Benchmarking is usually used to compare the information normally available in the public domain: products on the market, tools and practices, unit costs, resource efficiency, goals, published strategies, suppliers, for example.
"Benchmarking is not an afterthought of organizations that are highly skilled at strategic planning. It is not a one-time event to fulfill a reporting requirement of the budget cycle. Quite the contrary, benchmarking is a hallmark of effective strategy development. It is an on-going enabler of strategic design, strategic planning, and strategic thinking."-[Bogan, 1994, p. 178]

SUGGESTED IMPLEMENTATION:
- Know your competitors and follow their actions and products.
- Institute strategic benchmarking in enterprise.
- Train managers in legal benchmarking with cooperating competitors.
- Train all employees to "keep an open eye" to new inventions by competitors and report those to management.
- Allocate budget for industry fairs and subscribe to professional literature.

LAGGING FACTORS:
- Ignorance of who the competition is and what the competition is doing.
- Failure to schedule and budget for benchmarking.

SUGGESTED READING:
- C. Bogan, *Benchmarking Best Practices*, McGraw Hill, 1994.
- J.M. Morgan and J.K. Liker, *The Toyota Product Development System*. New York: Productivity Press, 2006, pp. 27–37, 339.

LEAN PRINCIPLE 5: Perfection

ENABLER: 3. **Use Lessons Learned from Past Programs for Future Programs.**

SUBENABLER(S) AND USE RANKINGS:
5. **Share metrics of supplier performance back to suppliers so they can improve.** (U = 0.39)

VALUE PROMOTED: *Right the first time* supplies, more predictable budget and schedule, reduction of frustrations between buyers and suppliers.

WASTE PREVENTED: All categories of waste.

EXPLANATION: As stated under several enablers, modern programs buy 60 to 95% or more of final value from vendors. Therefore, healthy relationships with suppliers are critical for success. "If we are going to be a benchmark supplier of equipment, our suppliers need to be benchmark suppliers of parts." (Anthony Pollock, Commodity Operations Manager at Xerox, [Bogan, 1994, p. 175]).

Even though suppliers may use their own metrics of performance, if buyers also measure suppliers' performance, these metrics should be shared with each supplier. The sharing indicates to suppliers where they are good and where improvements are needed in the eyes of the buyer. This tends to continually improve the buyer-supplier relationship and enables it to approach the asymptote of a seamless partnership. Popular metrics address quality, price, lead time, dependability, ability to operate and deliver JIT, cultural compatibility, responsiveness, and others. "Developing capable suppliers is of strategic importance for organizations such as Xerox and a growing number of other leading companies. With capable suppliers, an organization has a better chance of succeeding in its chosen markets." [Bogan, 1994, p. 175].

SUGGESTED IMPLEMENTATION:
- Implement a few well–thought-out key metrics of each supplier's performance. The metrics should measure those parameters that are important to the buyer's value creation and not be bureaucratic.
- Share the supplier performance with each supplier in the spirit of continuous improvement and partnership.
- Implement modern methods for supply chain management (hire an expert or send a manager to school for a graduate degree in supply chain management).

LAGGING FACTORS:
- *Hands-off* and *over-the-wall* relationships with suppliers.
- Selecting suppliers from the lowest bid.
- Blaming suppliers for buyer's own problems.

SUGGESTED READING:
- C. Bogan, *Benchmarking Best Practices*, McGraw Hill, 1994.
- J.M. Morgan and J.K. Liker, *The Toyota Product Development System*. New York: Productivity Press, 2006.

LEAN PRINCIPLE 5: Perfection

ENABLER: 4. Develop Perfect Communication, Coordination and Collaboration Policy across People and Processes.

SUBENABLER(S) AND USE RANKINGS:

1. **Develop a plan and train the entire program team in communications and coordination methods at the program beginning.** (U = 0.13)

3. **Promote good coordination and communications skills with training and mentoring.** (U = 0.33)

4. **Publish instructions for email distributions and electronic communications.** (U = −0.04)

6. **Publish a directory and organization chart of the entire program team and provide training to new hires on how to locate the needed nodes of knowledge.** (U = 0.38)

VALUE PROMOTED: Effective flow of work, reduced rework, *right the first time*.

WASTE PREVENTED: All categories of waste.

EXPLANATION: As described under subenabler 2.2.1, effective communications are critical to good Systems Engineering. Communications must be adequate to support program excellence, without consuming excessive time. As stated previously, engineers tend to lack effective communications skills. Companies and enterprises should develop appropriate communication policies and perform effective training. The present four subenablers promote the establishment of good policies for effective communications and coordination and for training. See Suggested Implementation for details.

SUGGESTED IMPLEMENTATION:
- Experts in effective engineering communications should develop instructions, the "dos and don'ts" for communications by all the means used in the enterprise: email, texting, phone, pagers, paper memos, stand-up meetings, *integrative events*, electronic social networks, and others, as applicable. For example, email instructions should say: "Send emails only to those individuals who need the information; do not include any attachments except an A3 summary form unless explicitly requested."
- As soon as an issue or risk is discovered, an A3 summary form should be sent to the indicated SE manager and possibly to the office of the Chief SE.
- Critical participants must be in attendance or on a teleconference, or the meeting will be a waste of time for the rest of the team. The perfect attendance must be made possible by detailed planning and scheduling of critical participants.
- Meetings must include a *Purpose, Agenda, Time, Intent* (PATI)

- Instructions for communications with suppliers should be issued.
- Pairs of employees should be designated across functions to serve as direct points of contact.
- A directory and organizational chart of all stakeholders for a given program should be developed online and kept current to minimize wasting time trying to find the right person. The program directory should list all relevant data for each person: name, email, location, job title, organization, office and cell phone numbers, pager, and other electronic locators, if used.
- All stakeholders should be trained in the use of A3 forms.
- Simple and user-friendly training materials should be issued, and team members trained.
- Train all new hires in the above communications.
- Since many of these instructions would benefit all programs, they should be implemented at the enterprise (corporate) level.

LAGGING FACTORS:
- Ad-hoc communications, lack of instructions and procedures.
- Culture of distributing emails "to everybody."
- Culture of verbose reports and memos drafted and redrafted many times and attached to massive emails
- Lack of knowledge how the best companies do it.

SUGGESTED READING:
- J.M. Morgan and J.K. Liker, *The Toyota Product Development System*. New York: Productivity Press, 2006.
- S. Hino, *Inside the Mind of Toyota*. New York: Productivity Press, 2006.
- B.W. Oppenheim, Lean product development flow. *Journal of SE*, 2004, **7**, 4, pp. 352–376.

LEAN PRINCIPLE 5: Perfection

ENABLER: 4. Develop Perfect Communication, Coordination, and Collaboration Policy across People and Processes.

SUBENABLER(S) AND USE RANKINGS:
2. Include communication competence among the desired skills during hiring. (U = 0.29)

VALUE PROMOTED: Effective communication among program stakeholders.

WASTE PREVENTED: All categories of waste.

EXPLANATION: An employee who possesses effective verbal and written communication skills "is worth his or her weight in gold." Yet, engineers and managers are often hired on the basis of technical skills, achievements, and experience, without regard for ability to communicate. The present subenabler promotes including a check of communication competence in the hiring process.

The practice of selecting candidates by computer scanning resumes for key words should be abandoned if practiced. Such selection ignores all the soft skills of a person, the energy, passion, enthusiasm, as well as nontechnical competencies, such as:

- Concise communication skills.

- Creativity (thinking outside the box).

- Teamwork.

- Ability to grasp situations quickly and thoroughly.

- Discipline to work consistently under strict timelines.

A person possessing these skills is likely to train, adopt, and perform better in many work situations. These skills are considered routinely when hiring at Toyota: "Each engineering applicant goes through a series of intensive interviews designed to provide a comprehensive look at personal characteristics that determine whether a prospective hire will fit within the Toyota culture." [Morgan and Liker, 2006, p. 169]

SUGGESTED IMPLEMENTATION: The importance of effective communication must be stressed in every program. A check of soft skills listed above should be routinely included in any process of hiring engineers and managers, but particularly so for Systems Engineers whose job includes coordination of the work of others. Such a test should be devised jointly by a small team comprised of both System Engineers and Human Resource managers. Examples follow:

- The request "Please tell me about yourself" forces the interviewee to demonstrate oral skills under stress.
- A request to write a sample memo on an assigned topic will reveal writing skills.
- A request to describe "preferred policies for effective communications and coordination" (or equivalent) reveals whether the candidate is aware of these critical needs.

LAGGING FACTORS:
- Good communications not valued in the enterprise.
- Hiring focused only on technical skills.
- Computer scanning of resumes.

SUGGESTED READING:
- J.M. Morgan and J.K. Liker, *The Toyota Product Development System*. New York: Productivity Press, 2006.
- J.H. Gittell, *The Southwest Airlines Way*, McGraw Hill, New York, 2003.

LEAN PRINCIPLE 5: Perfection

ENABLER: 4. Develop Perfect Communication, Coordination, and Collaboration Policy across People and Processes.

SUBENABLER(S) AND USE RANKINGS:
5. **Publish instructions for artifact content and data storage: central capture versus local storage, and for paper versus electronic, balancing between excessive bureaucracy and the need for traceability. (U = 0.33)**

VALUE PROMOTED: Orderliness, consistency, and traceability of documents, less bureaucracy—all of which translate into time savings and less rework.

WASTE PREVENTED: Waiting, inventory, rework, overprocessing.

EXPLANATION: Excellent record and artifact keeping is critical to technical traceability and avoidance of chaos and bureaucracy. Specific but easy-to-follow and user-friendly instructions should be issued at the beginning of program (or even better: at the enterprise level, since all programs will benefit from them) on the content and storage of artifacts. Specifics are suggested below.

SUGGESTED IMPLEMENTATION: Create and publish instructions for artifact content and data storage. The instructions should be user friendly, indicating easy-to-follow examples and steps, and include:
- Instruction which artifacts (versions) should be kept only in a central database and which can be kept on local media.
- What medium to use (paper versus electronic).
- How to release a document into a central database.
- Which documents need to be saved in a central location but not necessarily need to be formally released and which should go through the approval and release steps, with detailed instructions.
- Where the data is archived and how to search for information.
- Training and mentoring all stakeholders in the instructions.
- Establishing the discipline to follow instructions.
- Mentoring and offering good examples.

LAGGING FACTORS:
- Lack of effective processes and instructions for artifact content, release, and storage.
- Instructions written poorly, in a difficult language.
- Lack of training.
- Missed opportunities for technology reuse.

SUGGESTED READING:

- J.M. Morgan and J.K. Liker, *The Toyota Product Development System*. New York: Productivity Press, 2006.
- J.H. Gittell, *The Southwest Airlines Way*. New York: McGraw Hill, 2003.
- D. Secor, *Implementation of Lean Enablers for Systems Engineering at Rockwell Collins*, LAI Knowledge Exchange Event, LAI Plenary, March 23, 2010.

LEAN PRINCIPLE 5: Perfection

ENABLER: 4. Develop Perfect Communication, Coordination, and Collaboration Policy across People and Processes.

SUBENABLER(S) AND USE RANKINGS:
7. **Ensure timely and efficient access to centralized data. (U = 0.58)**
8. **Develop an effective body of knowledge that is historical, searchable, shared by team, and knowledge management strategy to enable the sharing of data and information within the enterprise. (U = 0.13)**

VALUE PROMOTED: A well-planned and designed central data storage is a great help to program stakeholders. The benefits include safety, security, backup, compatible formats, standardization, increased interoperability, better traceability, easy access to tradeoff curves and other technical design data, faster execution of technical tasks.

WASTE PREVENTED: All waste categories.

EXPLANATION: Data in central storage must be secure, protected against hacking and other unauthorized use, but it also must be easy and timely to find and use. (The author recalled an implementation where day-ahead application for access was required—which obviates the purpose.) Data should be available when needed, at a few clicks of a mouse.

The database should contain a body of knowledge based on as many past programs as is practical, comprised of important historical data, trade and trend curves, results from generic and envelope tests, prior solutions, modules of knowledge, etc. The data should be strategically designed and created at the enterprise level to benefit all programs in the enterprise.

Toyota's "V-comm" system is a good example of a successful *know-how* database. "This database has been successful for a number of reasons... the underlying company values, discipline, and reward for adhering to the process... so people automatically use and learn from this information. Moreover, the database is centralized; there are no competing knowledge databases to create confusion or contradictions. Functional specialists maintain, verify, and update their own portion of the database; just as they did with their handwritten checklists. ... It is a fully integrated network of tools that link active design, simulation, test, and communication technologies along with the *know-how* database to provide engineers with a powerful suite of tools to enhance their productivity. By integrating the *know-how* database into the basic engineering tool set, Toyota emphasizes that learning is not an add-on or 'extracurricular,' it is a basic part of the job" [Morgan and Liker, 2006, p. 283].

SUGGESTED IMPLEMENTATION: Implement a comprehensive central database that satisfies all needs, as well as assures data safety and security. This is a major strategic task for enterprises. This work should be executed by database experts in cooperation with functional managers, corporate managers, and IPT managers. It must start with a careful study of needs and benchmarks with other companies. This is a never-ending task because new knowledge is created all the time, and it must be captured in real time. Collaborative sharing of data among program customers and suppliers is another opportunity for creating effective knowledge base.

LAGGING FACTORS:
- No central database.
- Poorly designed database, with difficult access.
- New knowledge not captured in the central database.
- Lack of security and safety.
- Lack of training in the database use.

SUGGESTED READING:
- J.M. Morgan and J.K. Liker, *The Toyota Product Development System*. New York: Productivity Press, 2006.

LEAN PRINCIPLE 5: Perfection

ENABLER: 5. For Every Program Use a Chief Engineer Role to Lead and Integrate Development from Start to Finish.

SUBENABLER(S) AND USE RANKINGS:
1. The Chief Engineer role to be responsible, with authority and accountability for the program's technical success. (U = 0.48)
2. Have the Chief Engineer role lead both the product and people integration. (U = 0.04)
3. Have the Chief Engineer role lead through personal influence, technical *know-how*, and authority over product development decisions. (U = 0.17)
4. Groom an exceptional Chief Engineer role with the skills to lead the development, the people, and assure program success. (U = 0.04)
5. If Program Manager and Chief Engineer role are two separate individuals (required by contract or organizational practice), co-locate both to enable constant close coordination. (U = 0.29)

VALUE PROMOTED: Vastly better management of the program; RAA (responsibility, accountability, authority) focused in a single person.

WASTE PREVENTED: All categories of waste, including program failure or closure.

EXPLANATION: A frequent practice in recent U.S. governmental programs is to have two program managers: the *Program Manager* responsible for the program business success, and *Chief Systems Engineer* responsible for Systems Engineering. Numerous functional engineers are responsible for various technical areas. In some programs this causes split responsibilities, authorities, and accountabilities, often with imperfect results. "If everybody is responsible, nobody is responsible." In contrast, many U.S. and overseas commercial programs use only one person to be fully responsible for an entire program's success (both technical and business). This person is given various titles, Chief Engineer (very successful Toyota model, see Morgan and Liker [2006], Product Manager, Product Engineer, or similar). Early U.S. aerospace programs also used an extremely successful single-person *Chief Engineer* role (e.g., early Jack Northrop, Howard Hughes, Kelly Johnson of the Skunk Works, early NASA space programs, and others). Murman [2008] discusses some more recent successful programs with a single top manager in dual technical and business leadership roles. Subenabler 5.5.1 emphatically states that one and only one such person must be fully responsible, with authority and accountability for an entire technical program's success. The subenabler addresses only the technical role of the Chief Engineer, saying nothing about whether that person should also be the overall manager of the program or share the management with a separate business manager person. However, nothing in this subenabler should be taken as promoting the dual-head model. The dual-head model is not required under the U.S. government acquisition policies and is not promoted in the INCOSE Handbook version 3.2.

Subenablers 5.5.2–5.5.4 describe how to groom a Chief Engineer role if an enterprise presently lacks such competence.

SUGGESTED IMPLEMENTATION:
- Develop a long-term plan to groom the best individuals possessing the above characteristics for the Chief Engineer role. Use rotation, formal education, and increasing RAA over time.
- If a Program Manager is a separate person, co-locate him or her and the Chief Engineer role in adjacent offices for easy face-to-face communication and include in their evaluation the effectiveness of their joint decision making and cooperation.

LAGGING FACTORS:
- Dissolved technical management, no one really in charge.
- Financial program manager stronger than the technical manager(s).
- Program distributed among several contractors with poor central SE and poor coordination.

SUGGESTED READING:
- J.M. Morgan and J.K. Liker, *The Toyota Product Development System*. New York: Productivity Press, 2006.
- E.M. Murman, *Lean Aerospace Engineering*, Littlewood Lecture AIAA-2008-4, January 2008.
- A.C. Haggerty and E.M. Murman, *Evidence of Lean Engineering in Aircraft Programs*, 25th *International Congress of the Aeronautical Sciences,* Hamburg, Germany, September 2006.
- R. Leopold, *The Iridium Story: An Engineer's Eclectic Journey*, Minta Martin Lecture, MIT Department of Aeronautics and Astronautics, April 23, 2004.
- B.R. Rich and L. Janos, *Skunk Works*, Little & Brown, 1994.

LEAN PRINCIPLE 5: Perfection

ENABLER: 6. Drive out Waste through Design Standardization, Process Standardization, and Skill-Set Standardization.

SUBENABLER(S) AND USE RANKINGS:
1. Promote design standardization with engineering checklists, standard architecture, modularization, busses, and platforms. ($U = 0.57$)
2. Promote process standardization in development, management, and manufacturing. ($U = 0.67$)
3. Promote standardized skill sets with careful training and mentoring, rotations, strategic assignments, and assessments of competencies. ($U = -0.05$)

VALUE PROMOTED: *Right the first time* execution of tasks. Reduced variation to create highly stable and predictable outcomes.

WASTE PREVENTED: Rework, waiting, overprocessing, exceeded schedules and budgets.

EXPLANATION: Standardization plays a key role in Lean. The three subenablers promote three following aspects of standardization:
1. *Design standardization.* The philosophy of standardization and reuse of design modules (of parts, elements, components, software objects, subsystems, etc.) dramatically lowers the long-term cost of product development. For example, standardization makes automobile production affordable. Practically all subcomponents of cars are standardized before a new car is designed: engines, gearboxes, breaks, drive train elements, steering subsystem, dashboard electronics, computers, sensors, wheels, seats, etc. Standard modules are designed to be used in multiple products, tested, validated, and made robust. Then new car design is an efficient effort of trading off different modules to fit the available space, styling, weight, power, and performance requirements. With all modules predesigned, such trades can proceed efficiently and affordably (analogous to proverbial Lego blocks). Toyota designed the Prius car in nine months from the end of styling to the beginning of error-free production, even though the car introduced a new type of hybrid power plant. If these components were to be designed individually for each car model, cars would cost as much as satellites. Most high-tech products can follow this philosophy. Designing individual elements from scratch for each new product/program is the least efficient way of executing design effort. In government programs, contracts should be written to enable this long-term strategic reuse practice and abandon the focus on *one-off* strategies. The key tools for design standardization are engineering process checklists, reusable components, standard subsystems, and common architectures.

2. *Process standardization.* All repeatable processes should be standardized, so that they can be executed predictably, robustly, waste-free, ***right the first time***. Unpredictable processes potentially sabotage program planning: They make it impossible to estimate process effort, cost, and quality outcome. Standardization allows for synchronization of cross-functional processes that enable unprecedented product development speed. Standardizing a PD process means "standardizing common tasks, sequence of tasks and task effort and duration, and utilizing this as the basis for continuous improvement. Process standardization is a potent antidote to both task and inter-arrival variation. . . .It is how independent process/organizations know specifically what inputs are required from each other and when they are needed" [Morgan and Liker, 2006, p. 105]. Regretfully, many standard procedures suffer from difficult and unclear texts. They lack process optimization, and their release takes a long time with significant bureaucracy.

3. *Skill-set standardization.* Standardization of skills has been practiced in Lean manufacturing for many years with great benefits. The standards make it easy to assign employees to tasks with flexibility, use work cells, and organize Lean flow. Skill standardization among engineers is not much practiced in the Western world. Engineers tend to regard their work as unique, not subject to standardization. In fact, it is the work content that is unique, but most engineering processes should follow *best engineering practices*, which can and should be standardized with checklists, training and mentoring, rotations, strategic assignments, and assessments of competencies. Standard engineering skills may include the ability to operate different software tools; experience in executing checklists, analysis, design, or test; ability to interact with different stakeholders (e.g., customers and suppliers); and many others. Annual merit reviews should reflect the number of standard skills acquired during previous year. Department managers should motivate and incentivize engineers to complete new training modules. The department in which most employees are trained in most skills acquires great flexibility in assigning individuals to different IPTs dynamically and efficiently. "Toyota hires only about 1.1 percent of professional candidates applying for engineering positions and afterwards upholds a very rigorous process of maintaining skill proficiency as engineers gradually progress in their career. . . . A new engineer's career path consists of experiences that develop deep technical competence, while slowly climbing the technical hierarchy within each functional department, and is a direct result of engineers being rewarded for technical achievement." [Morgan and Liker, 2006, pp. 112–113]

SUGGESTED IMPLEMENTATION:

1. *Design standardization.* Strategic planning of design modules to standardize for future reuse should be done at the corporate level, based on long-term corporate strategy, trading off slightly higher initial costs versus long-term benefits of reuse. The work should be done by a trusted enterprise system architect. Once the decision is made, functions responsible for the modules should carry out the design, analysis, testing, and validation of the modules until they are ready for use.

2. *Process standardization.* Start with those processes that are executed most often and that are the most variable and frustrating. Employ applicable tools of process improvement: identify the sources of variation; perform a study of root cause and brainstorming, inviting process stakeholders; do not be afraid to *fail early-fail often*; optimize steps using a small team of best employee(s); and continue until variability and waste are eliminated. Then create a standard procedure, as follows:

   a. Draft procedure steps and test them on the least experienced employee(s).

   b. Redraft the procedure based on the comments and questions identified in (a).

   c. Finish the draft, make it clear, edit with visuals and examples.

   d. Release the new procedure efficiently, remembering that the procedure is to be a living document, efficiently updated with good new ideas and lessons learned.

   e. Offer a short training to all process stakeholders (often a few minutes during the weekly meeting is enough), saying "We have this new procedure for process X; the reason for issuing/updating it is [...]; here are the critical dos and don'ts; you find the procedure in [location]."

   f. Continue monitoring the procedure use and outcomes and mentor the employees.

3. *Skill set standardization.* Categorize all repeated work in the department into standard skill sets. Develop, prioritize, and implement standard training for each skill set. Incentivize employees to complete the training modules. Monitor and mentor individuals. Promote rotations, strategic assignments, and assessments of standard competencies.

LAGGING FACTORS:

- Never enough time to improve the work or processes.
- The conviction that all work is unique and nothing is repeatable.
- Ad-hoc management of human resources.
- Lack of long-term corporate strategy for product line.

SUGGESTED READING:

- J.M. Morgan and J.K. Liker, *The Toyota Product Development System*. New York: Productivity Press, 2006.

LEAN PRINCIPLE 5: Perfection

ENABLER: 7. Promote All Three Complementary Continuous Improvement Methods to Draw Best Energy and Creativity from All Employees.

SUBENABLER(S) AND USE RANKINGS:

1. **Utilize and reward bottom-up suggestions for solving employee-level problems. (U = 0.17)**

2. **Use quick response small *Kaizen* teams comprised of problem stakeholders for local problems and development of standards. (U = 0.13)**

3. **Use the formal large Six Sigma teams for the problems that cannot be addressed by the bottom-up and *Kaizen* improvement systems, and do not let the Six Sigma program destroy those systems. (U = 0.13)**

VALUE PROMOTED: Predictable robust flow of value adding work.

WASTE PREVENTED: All categories of waste.

EXPLANATION: Organizations must have effective means of preventing and mitigating all problems, frustrations, wastes, complaints, and so on. For competitive reasons and for the internal health of the organization, continuous improvement (CI) must be conducted for all activities (subject to reasonable limits on overprocessing waste, defined by subenablers 5.2.6 and 5.2.7). The spirit that "we can always do this better" must be present. The range of popular CI methods can be categorized as follows:

a. Top-down directives from management. Called "theory X management" (only managers know best), still preferred by some managers, this tool is the least effective: It promotes authoritarian management style. Using this CI method, the organization has at best as many improvers as is the number of managers—never enough to see and improve everything. Typically, managers are too busy with everyday work to invest much time into improvements. (The exception is a strong support for continuous improvement—increasing customer value and eliminating waste—at the CEO level. This leadership is critically needed). Employees not empowered to suggest improvements tend to think of frustrations, grievances, job jumping. ... Far more effective are the following continuous improvement methods.

b. Bottom up suggestions from employees for improvements in their own area of work. Toyota sets a good example for empowering and incentivizing all employees to think of improvements and bring them to the attention of supervisors, who tend to encourage the implementation. Bonuses are paid for implemented ideas. This powerful practice results in as many potential *improvers* as the number of employees. In the NUMMI (New United Motor Manufacturing, Inc.), California, plant 90% of employees submitted improvement suggestions, of which 60% were implemented. The employees engaged in improving their work experience take pride in their work, and identify more with their employer.

c. Small *Kaizen* quick-reaction teams of local stakeholders for slightly larger problems involving several stakeholders.

Kaizen typically requires a few people (two to five local stakeholders) working for a few hours or days to address an urgent local problem: find root cause, experiment with the best improvement steps, draft a new procedure, improve the time or effort of a task, prevent a recurring frustration, make the process predictable, etc. *Kaizen* teams are best suited for the cross-functional problem of size larger than what can be handled by one employee. *Kaizen* relies on enthusiasm of team members working with minimum formalities. *Kaizen* requires management support. Tangible recognition should be offered for success.

d. The *big bang* method of Six Sigma for big problems.

The term Six Sigma is adopted from statistics of production defects. Inspired by tightening tolerance requirements in modern electronics, Six Sigma has evolved to be a disciplined and structured company-wide effort to fix large problems in all areas of corporate activity. Individuals are chosen and trained for days or weeks in improvement methods. As full- or part-time Six Sigma experts, the chosen individuals work on dedicated Six Sigma projects. The projects—typically large, taking weeks or months to complete—are prioritized from highest to lowest importance using formal quality statistics. Six Sigma is a disciplined method of rigorous data-gathering and analysis to pinpoint sources of errors and ways of eliminating them, with heavy reliance on performance metrics.

Harry and Schroeder [2000] describe Six Sigma as being "a business process that allows companies to drastically improve their bottom line by designing and monitoring everyday business activities in ways that minimize waste and resources while increasing customer satisfaction. Six Sigma guides companies into making fewer mistakes in everything they do—from filling out a purchase order to manufacturing airplane engines—eliminating lapses in quality at the earliest possible occurrences."

In practice, Six Sigma demonstrates its effectiveness through improving large problems that require formal statistical studies of causes of variation and problems, followed by formal corrective processes. Unfortunately, many companies have implemented Six Sigma in a bureaucratic and wasteful way. Six Sigma tends to be an overkill for small local problems that are best addressed by individual employees or small *Kaizen* teams. Regretfully, some companies decided that Six Sigma can serve as the sole CI method, and they eliminated (b) and (c) in the process. The present subenablers strongly promote continued use of (b) and (c) for small local problems, and the use of (d) for truly large problems.

SUGGESTED IMPLEMENTATION: Engage frontline employees to become problem solvers. Promote all three CI methods: (b), (c) and (d), and use (a) sparingly.

(b) requires a short training, followed by mentoring and practice, and a lot of encouragement. It can be implemented by individual managers in their area of work. The issue of bonuses or honorary rewards should be addressed.

(c) requires a short training, followed by mentoring and practice, and can be implemented by individual managers in their area of work.

(d) is a corporate-wide initiative that should be implemented by experts at high management levels and will require significant corporate resources.

LAGGING FACTORS:
- Authoritarian management style, lack of empowerment of employees.
- Bureaucratic implementation of Six Sigma alone.

SUGGESTED READING:
- W.E. Deming, *Out of the Crisis*. Massachusetts Institute of Technology, Center for Advanced Engineering Study, 1982.
- HBR., Ed., *Manufacturing Excellence at Toyota*, Harvard Business School Series, 2008.
- J.M. Morgan and J.K. Liker, *The Toyota Product Development System*. New York: Productivity Press, 2006.
- M. Harry and R. Schroeder, *Six Sigma: The Breakthrough Management Strategy Revolutionizing The World's Top Corporations*. Currency Doubleday, 2000.

7.2.6   Lean Principle 6: Respect for People

Summary   The five enablers listed here address several important aspects of human relations at work, as follows. The numbers in parentheses list the number of subenablers under each enabler.

Enabler 6.1 Pursue People Management According to the INCOSE Handbook Process.

The first enabler listed under each principle emphasizes the same fundamental point: to add the wisdom of Lean Thinking to the traditional Systems Engineering process rather than replace the process. So, again, the present enabler recommends to continue using the INCOSE SE Handbook, or any other equivalent SE handbook or manual, and in addition to implement the enablers that follow.

Enabler 6.2 Build an Organization Based on Respect for People. (14)

The 14 subenablers include a comprehensive checklist addressing leadership practices. They speak to constructive human relations among stakeholders; good practices for selecting and promoting employees; employee empowerment; and *flow down* of responsibility, authority, and accountability to the lowest employee level.

Enabler 6.3 Expect and Support Engineers to Strive for Technical Excellence. (4)

The four subenablers promote the establishment of communities of practice and investment in workforce development and Lean training.

Enabler 6.4 Nurture a Learning Environment. (8)

The eight subenablers present a comprehensive checklist for creating a learning environment, including training, continuing education, and providing knowledge experts to stakeholders.

Enabler 6.5 Treat People as Most Valued Assets, Not as Commodities. (0)

This self-explanatory enabler, with no subenablers, adopted from the teachings of Deming [1982], has been chosen to serve as a closure for the entire set because of its profound importance to Systems Engineering. It is as valid today as it was in the time of that respected expert on organizations.

The following tables describe the individual subenablers.

LEAN PRINCIPLE 6: Respect for People

ENABLER: 2. Build an Organization Based on Respect for People.

SUBENABLER(S) AND USE RANKINGS:
1. Create a vision that draws and inspires the best people. (U = 0.58)

VALUE PROMOTED: Extraordinary systems can be created by an extraordinary team of professionals who become inspired by a great leader.

WASTE PREVENTED: All categories of waste.

EXPLANATION: History is rich with examples of extraordinarily successful human endeavors led by visionary leaders. Some examples include Skunk Works, led by Kelly Johnson and later Ben Rich; early Northrop company, led by Jack Northrop; U.S. nuclear submarine program, led by Admiral Rickover; Toyota Prius, led by Akihiko Otsuka; early Microsoft, led by Bill Gates; the Apple I-products, led by Steve Jobs; and numerous others. All these programs were led by strong leaders who projected inspiring visions, manifested high levels of competence, and were able to build amazing teams of professionals to create extraordinary products. The best companies place such leaders at the top of the corporate hierarchy and reward accordingly (in contrast to the present appalling trend of offering astronomical salaries to CEOs who focus only on short-term financial gain.) Many recent defense programs have departed strongly from those early industrial successes and are now characterized by impersonal management dissolved among many individuals, and organizations and technical leadership replaced by bureaucrats counting artifacts.

SUGGESTED IMPLEMENTATION: Companies should make every effort to find, hire, groom, and treasure the best technical leaders they can find. In the corporate pecking order, a good technical leader should be valued and rewarded higher than managers who are only focused on short-term financial metrics. Once a leader is hired, he or she should receive maximum institutional support. Government should support technical leadership.

LAGGING FACTORS: Bureaucratic, leaderless, dissolved management focused on short-term financial performance.

SUGGESTED READING:
- J.M. Morgan and J.K. Liker, *The Toyota Product Development System*. New York: Productivity Press, 2006.
- E.M. Murman, *Lean Aerospace Engineering*, Littlewood Lecture AIAA-2008-4, January 2008.
- R. Leopold, *The Iridium Story: An Engineer's Eclectic Journey*, Minta Martin Lecture, MIT Department of Aeronautics and Astronautics, April 23, 2004.
- J.H. Gittell, *The Southwest Airlines Way*, New York: McGraw Hill, 2003.
- B.R. Rich and L. Janos, *Skunk Works*. Little & Brown, 1994.

LEAN PRINCIPLE 6: Respect for People
ENABLER: 2. Build an Organization Based on Respect for People.
SUBENABLER(S) AND USE RANKINGS: 2. **Invest in people selection and development to promote enterprise and program excellence.** (U = **0.46**)
VALUE PROMOTED: Great employees are the best means for creating best value and preventing all categories of waste.
WASTE PREVENTED: All categories of waste.
EXPLANATION: This subenabler emphasizes the importance of seeking highly competent and motivated employees that will benefit a company in the long run. Most companies examine a candidate's skills and educational background to see if he or she will be a good fit. However, great companies go farther: "In a lean PD system, being an engineer is a calling rather than a job" [Morgan and Liker, 2006, pp. 223–224].
SUGGESTED IMPLEMENTATION: Hire employees for long-term employment, based on their entire worth: not only narrow professional skills and experience, but also teamwork abilities, learning, communication, adaptability, and leadership skills, or potential. Then, invest in employee development: train and rotate employees to become acculturated to the company's best habits and processes, becoming enthusiastic members and leaders of their teams.
LAGGING FACTORS: • Hiring for an immediate short-term technical need. • Hiring based on scanning of resumes for key words. • Expediency. • Treating employees as commodities.
SUGGESTED READING: • J.M. Morgan and J.K. Liker, *The Toyota Product Development System*. New York: Productivity Press, 2006. • R. Leopold, *The Iridium Story: An Engineer's Eclectic Journey*, Minta Martin Lecture, MIT Department of Aeronautics and Astronautics, April 23, 2004. • J.H. Gittell, *The Southwest Airlines Way*, New York: McGraw Hill, 2003.

LEAN PRINCIPLE 6: Respect for People
ENABLER: **2. Build an Organization Based on Respect for People**.
SUBENABLER(S) AND USE RANKINGS: **3. Promote excellent human relations: trust, respect, honesty, empowerment, teamwork, stability, motivation, drive for excellence**. **(U = 0.71)**
VALUE PROMOTED: Vastly better teamwork skills, enthusiasm for work, creativity, energy, focus.
WASTE PREVENTED: All categories of waste.
EXPLANATION: "Respect for people" is one of the pillars of the successful Toyota culture, best summarized by Deming [1982, p. 148]: "Once a Japanese plant manager who turned an unproductive U.S. factory into a profitable venture in less than three months told me that: 'It is simple. You treat American workers like human beings with ordinary human needs and values, they react like human beings.' Once the superficial, adversarial relationship between managers and workers is eliminated, they are more likely to pull together during difficult times and to defend common interests in the firm's health."

The individual aspects listed in subenabler 6.2.3 carry rich practical meanings, as follows:

Trust: Each employee must trust his or her colleague in any transaction. Each knows that the colleague making a request or answering a question does so for a legitimate reason and not for his or her own personal benefit. Both individuals must trust each other's intent to maximize value while minimizing waste. In case of conflict, they should try to negotiate together with focus on customer value.

Respect: People who are respected tend to respect others, which is conducive to a healthy culture focused on value, creativity, and continuous improvement. In contrast, people working in a culture lacking mutual respect tend to think mostly of grievances and frustrations.

Honesty and openness: The parties in any communication must trust each other that the information provided represents an honest and open assessment based on available facts and professional judgments and is not driven by private interests. The openness means that all stakeholders receive the same message.

Empowerment: The employees who feel empowered to make decisions and solve problems at the lowest level, and to resolve conflicts at the lowest level, tend to utilize their creativity and responsibility more than those who feel disciplined to carry out only direct orders of managers.

Teamwork: Practically all employees and stakeholders involved in modern PD programs work in teams. The ability to work effectively on a team and reach consensus is critical for success.

Stability: In complex programs the time needed to become an effective team member experienced in the domain and company culture takes years and requires dedication. Employees should be given an environment conducive to focus on work, and job stability is a critical element of that environment. In contrast, amid fear of layoffs, employee energy tends to shift toward finding other work.

Motivation: In a normal system, except in rare cases (such as substance abuse, excessive absences, and criminal acts) that, according to Deming [1982] constitute only a small percentage of the workforce, the vast majority of employees at all levels want to work well. The role of an employer is to provide employees with a work environment that is based on respect, empowerment, trust, good communications, teamwork, mutual support, lack of internal rivalry, fair evaluations, and the sense that value in the organization is being pursued and waste minimized at all times.

Drive for excellence: Again, except for rare cases, pursuit of excellence is a powerful human desire. Employees should be given a chance and encouragement to pursue excellence, and be rewarded for doing so.

SUGGESTED IMPLEMENTATION: Implementation of great work culture is said to be the most difficult aspect of management. It requires:

- Long-term thinking and long-term efforts.
- True leadership from enterprise leaders.
- Tangible policy to disseminate the best practices throughout the enterprise.
- A policy of hiring based not only on technical competence but also on excellent interpersonal skills.
- Frequent and effective mentoring at all levels of management.
- Formal training and exposure to best examples.
- Mentoring.

LAGGING FACTORS:

- Authoritarian management.
- Excessive focus on short-term financial performance rather than long-term competitiveness.
- Bad hiring practices (focused only on technical skills or minimum wages).
- Lack of leadership.
- Unfair compensation policies.
- Adversarial relationships between unions and management.
- Lack of job stability, frequent layoffs.

SUGGESTED READING:

- David Levine, *Reinventing the Workplace: How Business and Employees Can Both Win*, Brookings Institution Press, 1995.
- J.K. Liker, *The Toyota Way: 14 Management Principles from the World's Greatest Manufacturer*. McGraw-Hill Professional, 2004.
- J.M. Morgan and J.K. Liker, *The Toyota Product Development System*. New York: Productivity Press, 2006.
- R. Leopold, *The Iridium Story: An Engineer's Eclectic Journey*, Minta Martin Lecture, MIT Department of Aeronautics and Astronautics, April 23, 2004.
- J.H. Gittell, *The Southwest Airlines Way*, New York: McGraw Hill, 2003.

LEAN PRINCIPLE 6: **Respect for People**
ENABLER: **2. Build an Organization Based on Respect for People**.
SUBENABLER(S) AND USE RANKINGS: **4. Read applicant's resume carefully for both technical and nontechnical skills and do not allow mindless computer scanning for keywords.** **(U = 0.50)**
VALUE PROMOTED: Best competitive products require best teams with best employees hired for a long-term employment.
WASTE PREVENTED: All categories of waste.
EXPLANATION: A sad recent practice in large corporations is to filter job applicants' resumes using computer scanning for keywords. This is done to save a few jobs in the Human Resource Department, a dramatic example of shortsighted cost cutting. All human beings, even at entry level, are complex individuals with rich biographies, experiences, accomplishments, interests, creativity levels, leadership skills, temperament, and passion. It is precisely these intangible features rather than the mechanical side of the professional knowledge that separate mundane workers from creative enthusiasts and leaders. This scanning practice ignores all these wonderful human aspects and focuses on the technical keywords, selecting people as commodities rather than assets. Toyota is a classical example of hiring for lifetime employment: "This is why Toyota puts such a tremendous effort in finding and screening prospective employees. It wants the right individuals to train and empower to work in teams. When Toyota selects one person out of hundreds of job applicants after searching for many months, it is sending a message—the capabilities and characteristics of individuals matter" [Liker, 2004, p. 186].
SUGGESTED IMPLEMENTATION: If computer scanning for keywords is practiced, abandon it immediately. Train the Human Resource employees who filter resumes to pay attention not only to professional qualifications, but also, and with emphasis, to the above intangible human aspects. Working with experts, develop tests for identifying the best human characteristics, to be sought during hiring process.
LAGGING FACTORS: • Hiring based on resume scanning for keywords. • Hiring for an immediate short-term need. • Treating employees as commodities.

SUGGESTED READING:

- J.K. Liker, *The Toyota Way: 14 Management Principles from the World's Greatest Manufacturer*. McGraw-Hill Professional, 2004.
- J.M. Morgan and J.K. Liker, *The Toyota Product Development System*. New York: Productivity Press, 2006.
- R. Leopold, *The Iridium Story: An Engineer's Eclectic Journey*, Minta Martin Lecture, MIT Department of Aeronautics and Astronautics, April 23, 2004.
- J.H. Gittell, *The Southwest Airlines Way*, New York: McGraw Hill, 2003.

LEAN PRINCIPLE 6: Respect for People

ENABLER: 2. Build an Organization Based on Respect for People.

SUBENABLER(S) AND USE RANKINGS:
5. Promote direct human communication. (U = 0.63)

VALUE PROMOTED: Effective creation and transmittal of vast amount of information needed in programs.

WASTE PREVENTED: Primarily rework, but all other categories of waste, too.

EXPLANATION: Product development, especially systems engineering, involves creation and flow of information among project stakeholders. "Today, if you ask a room full of engineers if communication is important to effective product development, you are likely to get amused shrugs—the answer is obvious. After all, product development is information flow among many specialists. Stop communication, stop information flow, and you stop product development. Now, instead of 'throwing the design *over-the-wall*,' (emphasis by the present author) engineers are taught to communicate concurrently with a team of upstream and downstream specialists—across functions. Almost everyone understands that more communication is better and that collocating engineers in the same office area so that they communicate intensely every day is a PD 'best practice'."[Morgan and Liker, 2006, p. 259]

Communication should be conducted as directly as possible, in real time, efficiently, and with minimum intermediate handoffs. Both parties to a communication should feel obligated to make certain that their communication is clear, complete, and efficient.

When an engineer needs to find an answer to his or her question from another engineer working in a different department, he or she should seek a direct contact rather than go through managers. Furthermore, managers should empower such direct contact. Communication should utilize the broadest bandwidth available: face to face is better than video link, which is better than phone, which is better than email, which is better than using some automated impersonal management software. Employees must be trusted (and developed to be worthy of the trust) to communicate verbally and effectively.

Similarly, communications between buyer and supplier engineers should be direct; however, subject to certain reasonable legal limitations to prevent *requirements creep* and specification changes, as explained under subenabler 3.5.3.

SUGGESTED IMPLEMENTATION: On average, engineers are known to be rather poor communicators and prefer the language of mathematics, computers, and images to spoken words, which is often why they selected engineering as their profession. (Computer scientists are said to be particularly notorious for preferring computer over human communications.) Yet, in complex programs, verbal communication and coordination are critical to program success. It follows that a key role of management is to mentor, promote, expect, and enforce effective human communications. Management should hire and promote individuals based not only on their technical skills but also on human communication skills.

While to some degree good human communication skills may be a natural gift, to a large degree the skills can be improved by focused effort, practice, mentoring, and good examples. Management should organize ample opportunities.

LAGGING FACTORS:
- Cultures that promote automated impersonal communications.
- Managers who themselves are poor communicators.
- Managers who demand that "everything goes through me."

SUGGESTED READING:
- J.M. Morgan and J.K. Liker, *The Toyota Product Development System*. New York: Productivity Press, 2006.
- W.E. Deming, *Out of the Crisis*. Massachusetts Institute of Technology, Center for Advanced Engineering Study, 1982.
- R. Leopold, *The Iridium Story: An Engineer's Eclectic Journey*, Minta Martin Lecture, MIT Department of Aeronautics and Astronautics, April 23, 2004.

LEAN PRINCIPLE 6: Respect for People
ENABLER: **2. Build an Organization Based on Respect for People**.
SUBENABLER(S) AND USE RANKINGS: **6. Promote and honor technical meritocracy**. (U = **0.83**)
VALUE PROMOTED: Better motivation to enter and continue in the engineering profession, larger pool of engineers available for PD and SE, healthier human relations at work.
WASTE PREVENTED: All categories of waste.
EXPLANATION: In 1983, journalist Edward A. Reynolds lamented the fact that the corporate focus has shifted from technical to financial: "Practically, all of our major corporations were started by technical men—inventors, mechanics, engineers, and chemists, who had a sincere interest in quality of products. Now these companies are largely run by men interested in profit, not product. Their pride is in the P & L statement or stock report." [*Standardization News*, Philadelphia, April 1983, p. 7]. Edward Deming, the world guru of quality, included the above quote in his milestone book *Out of the Crisis* [1982, p.131]. Since that time, the situation has gotten worse. Wall Street cravings for short-term dividends has destroyed long-term investment and R&D in many U.S. high-tech corporations and placed financial managers well above engineers on all scales: compensation, job perks, and prestige. Engineering is losing its esteem. The fraction of U.S. high school graduates who enter engineering schools, already vastly smaller than in other leading countries, is steadily dropping. The number of U.S. born and naturalized engineers who can obtain security clearance is also decreasing every year. As a result, defense companies are increasingly outsourcing engineering work abroad. In many industries, including defense, long-term corporate-sponsored research has disappeared, regarded as taking away from dividends. The present subenabler promotes a broad reversal of these trends. After all, the products being designed, built, and used are still satellites, aircraft, ships, cars, computers, and other such goods, and not stock reports. The companies making these goods can remain competitive only if their products are competitive. In order to make them so in the long term, intensive long-term R&D is needed. Excellent creative technical people are needed to carry out R&D and their subsequent engineering development programs. Therefore, this subenabler should be interpreted to mean: • Promote based on technical meritocracy and not financial or political expediency. • Improve professional standing of engineers relative to financial managers on all scales: compensation, prestige, perks.

- Promote long-term focus on internal R&D, even if short-term dividends suffer.
- Make every effort to make the engineering profession attractive to high school graduates.
- Government must also strongly promote long term R&D.

Again, Toyota is a good example: The Chief Engineer at Toyota who is in charge of a new car PD is a person with equal prestige and compensation to a corporate vice president. Everybody knows him and respects him. People refer to the car he develops by his name rather than by the car name.

SUGGESTED IMPLEMENTATION:
- The corporate boards of high-tech firms must recognize that without long-term R&D there is no long-term future.
- In order to develop R&D, the balance between financial and technical corporate forces must change in favor of technical.
- In order to make the engineering profession more attractive, engineers must be recognized at least as highly as financial managers in prestige and compensation.
- The government acquisition policy must support long-term R&D in high-tech companies.
- Engineers must be promoted based on technical merit.

LAGGING FACTORS:
- Financial pressures to maximize short-term profits.
- Government pressures to pay only for PD programs and not for any long-term R&D.
- Corporate policy of valuing financial managers higher than engineers.

SUGGESTED READING:
- J.M. Morgan and J.K. Liker, *The Toyota Product Development System*. New York: Productivity Press, 2006.
- W.E. Deming, *Out of the Crisis*. Massachusetts Institute of Technology, Center for Advanced Engineering Study, 1982.

LEAN PRINCIPLE 6: Respect for People

ENABLER: 2. Build an Organization Based on Respect for People.

SUBENABLER(S) AND USE RANKINGS:
7. Reward based upon team performance and include teaming ability among the criteria for hiring and promotion. (U = 0.25)

VALUE PROMOTED: Better PD team is conducive to better creation of value with less waste in faster time.

WASTE PREVENTED: All categories of waste.

EXPLANATION: A leftover of the discredited "X-theory" of management, rivalry between workers on the same team is destructive. Competition is healthy among competing teams but not among players on the same team. We tend to understand it well in the context of sports but not work: Imagine a mountain-climbing team in which climbers compete for ropes and hooks instead of securing one another.

All employees in an enterprise should work together as a team aligned to create value with minimum waste in the fastest possible time. Teamwork means that more experienced individuals mentor and help those less experienced, because enterprise wins if the team as a whole performs better than its competitors.

This subenabler promotes two practices, as follows. First, the teaming ability should play an important part among the criteria for hiring and promotion. Second, an employee bonus should be based, at least partly, on the entire team performance rather than individual performance.

SUGGESTED IMPLEMENTATION:
- Include teaming ability among the criteria for hiring and promotion. Create special questions and tests to be used during the hiring interview process to detect the ability or lack thereof. Include teaming ability in annual evaluations.
- Reward based upon team performance. Make the total compensation consist of three components: base pay commensurate with education, experience and position, plus a bonus (identical for all team members) dependent on the entire team performance, plus an individual bonus for continuous improvement.

LAGGING FACTORS:
- The "X-theory" management, lack of teamwork.
- Rivalry among individual team members.
- Hiring only for technical skills.

SUGGESTED READING:

- J.K. Liker, *The Toyota Way: 14 Management Principles from the World's Greatest Manufacturer*. McGraw-Hill Professional, 2004.

- J.M. Morgan and J.K. Liker, *The Toyota Product Development System*. New York: Productivity Press, 2006.

- R. Leopold, *The Iridium Story: An Engineer's Eclectic Journey*, Minta Martin Lecture, MIT Department of Aeronautics and Astronautics, April 23, 2004.

- J.H. Gittell, *The Southwest Airlines Way*, McGraw Hill, New York, 2003.

LEAN PRINCIPLE 6: Respect for People

ENABLER: 2. Build an Organization Based on Respect for People.

SUBENABLER(S) AND USE RANKINGS:

8. *Flow down* responsibility, authority, and accountability (RAA) to allow decision making at lowest appropriate level. (U = 0.09)

9. Eliminate fear and promote conflict resolution at the lowest level. (U = 0.29)

12. Within program policy and within their area of work, empower people to accept responsibility by promoting the motto "ask for forgiveness rather than ask for permission." (U = 0.28)

VALUE PROMOTED: Free of fear, well-motivated, and empowered workforce tends to perform better in value creation with minimum waste.

WASTE PREVENTED: All categories of waste.

EXPLANATION: At its most basic, there are two competing management cultures. The first is known as "traditional," or "Theory X." It is the authoritarian hierarchical structure, in which the focus of power is vested in the hands of a superior who makes all decisions and commands obedience. In this culture vertical organizational *stovepipes* are common, and fear is the frequent denominator. Fighting it, [Deming 1982, p. 264] formulated his famous 8th Principle: "Drive out fear. No one can put in his best performance unless he feels secure. *Se* comes from the Latin, meaning without, *cure* means fear or care. *Secure* means without fear, not afraid to express ideas, not afraid to ask questions. Fear takes on many faces. A common denominator of fear in any form, anywhere, is loss from impaired performance and padded figures."

The opposite culture, sometimes called "Theory Y" management, is based on empowerment; delegation of responsibility, authority, and accountability; teamwork; and respect for all. Instead of vertical *stovepipes*, emphasis is placed on the creation and horizontal flow of value across all tasks, all the way to an end customer. Line managers make certain that employees have the right resources, including training and mentoring, to create value with minimum waste, and then happily delegate and empower employees to makes decisions and resolve conflicts themselves, consistent with value proposition and the corporate mission.

SUGGESTED IMPLEMENTATION: Transforming the enterprise with a traditional management culture to one of empowerment may not be easy, but this has been done numerous times. It requires committed leadership, a lot of training, and mentoring from the top. The most dramatic example of success was the NUMMI plant, a joint venture of Toyota and GM in Fremont, California, originally a GM plant. After GM closed the plant, blaming the workers and the UAW union for terrible plant performance, the new joint venture reopened under Toyota management, with the same workers and the same union. Massive training of all workers in the Toyota Production System and Respect for People followed. Within a year of opening, the NUMMI plant was transformed with extraordinary results, praised equally by the new management and the old union.[*] The joint venture published an amazing union agreement, which is now used as a classical case study in most business schools (see the reading list). Before making any layoffs, it promises a cut in salaries of executives and officers, insourcing work, and voluntary leaves. It also promises "commitment to building and maintaining the most innovative and harmonious labor-management relationship in America . . . .mutual trust, understanding and sincerity, and [. . .] work environment [. . .] based on teamwork, mutual trust and respect that gives recognition to the axiom that people are the most important resource of the Company."

This dramatic example is the case of the proverbial *burning platform* after the original GM plant closed. A less dramatic implementation may be to start with one *worst* department. A new manager who is a believer in Lean and has a strong track record of implementing the new management style should be carefully hired to transform this department. Then, after perhaps a year, when success is evident, the department can be used as an example for others. Good word spreads like fire: Employees from other departments will also desire the new culture.

---

[*]The NUMMI closed again in 2010 after bankrupt GM pulled out from the joint venture. Toyota then decided that the plant infrastructure was too large and costly to operate alone.

LAGGING FACTORS:
- Authoritarian management.
- Culture of fear.
- Lack of leadership.
- Focus on profits rather than value creation.

SUGGESTED READING:
- *Agreement between New United Motor Manufacturing, Inc. and the United Auto Workers*, August 1, 1998.
- David Levine, *Reinventing the Workplace: How Business and Employees Can Both Win*, Brookings Institution Press, 1995.
- J.K. Liker, *The Toyota Way: 14 Management Principles from the World's Greatest Manufacturer*. McGraw-Hill Professional, 2004.
- J.M. Morgan and J.K. Liker, *The Toyota Product Development System*. New York: Productivity Press, 2006.
- W.E. Deming, *Out of the Crisis*. Massachusetts Institute of Technology, Center for Advanced Engineering Study, 1982.

LEAN PRINCIPLE 6: Respect for People
ENABLER: 2. Build an Organization Based on Respect for People.
SUBENABLER(S) AND USE RANKINGS: 10. Keep management decisions crystal clear but also promote and reward the bottom-up culture of continuous improvement and human creativity and entrepreneurship. (U = 0.04)
VALUE PROMOTED: The company that is continuously improved by all employees becomes more competitive faster.
WASTE PREVENTED: All categories of waste.
EXPLANATION: This subenabler expands on the subenablers 6.2.8, 6.2.9, and 6.2.12, promoting the culture of empowerment, delegation, and *flowing down* of RAA (responsibility, accountability, and authority). The present subenabler (6.2.10) promotes bottom-up suggestions for continuous improvement. In this culture we have as many potential improvers as are employees. Employees themselves can devote their energies, creativity, and entrepreneurship to think of improvements in value creation, work conditions, and waste reduction. In contrast, in the traditional authoritarian management culture we have at best as many improvers as managers—and even this is often a fallacy because such managers tend to be so overwhelmed trying to control all activities and crises that they rarely have time to think of improvements. Such organizations tend to be very static and move from crisis to crisis. Several quotes illustrate the bottom-up culture: • "Until senior management gets their egos out of the way and goes to the whole team and leads them all together … senior management will continue to miss out on the brain power and extraordinary capabilities of all their employees. At Toyota, we simply place the highest value on our team members and do the best we can to listen to them and incorporate their ideas into our planning process." Alex Warren, former Senior VP Toyota Motor Manufacturing, Kentucky [Liker, 2004, 171] • "Under the suggestion program [employees] come up with work improvement ideas and they are rewarded by the company for good suggestions. This system has become the most common of the employee motivation systems and is used at 93.9% of all manufacturing companies listed on the Tokyo Stock Exchange." [Mikami, 1982, pp. 59–60]

- "According to surveys on the results of suggestion programs conducted in 1981 by the Japan Human Relations Association, 1.83 million employees at the 464 companies that responded submitted a total of 23.53 million suggestions. Hitachi, Ltd. Reported 4.21 million suggestions, which was the largest number for any company. The total amount awarded for the suggestions at Hitachi was 8.1 billion yen (466 yen per suggestion). The company confirmed that the economic effect was 225.3 billion yen. The suggestions were for:

  a. Improved work methods   34%

  b. Conservation of resources, materials, and consumables   13%

  c. Improved work environment   11%

Most of these suggestions were concerned with improvements related to individual work." [Mikami, 1982, p. 60]

The bottom-up suggestion system originated in manufacturing, but nothing should prevent its implementation in SE and PD environments.

SUGGESTED IMPLEMENTATION: The literature on the NUMMI plant conversion provides details of a plant-wide implementation of the bottom-up improvement system, which took about a year. The implementation is easier if started within a single department by a manager who is enthusiastic about the culture change. A short briefing explaining the suggestion system is a good place to start. Incentives in the form of small bonuses, or at least honest honorary recognition, are recommended. In order to motivate supervisors to accept suggestions positively, a small fraction of the bonus paid to the employee should also be paid to the supervisor (Toyota pays 5% of the bonus to supervisors automatically, whenever a suggestion is implemented). Overtime, and other reasonable resources should be made available to implement such suggestions. Once an idea catches on, other departments will likely start pulling for similar changes.

LAGGING FACTORS:
- Authoritarian, fear-based management.
- *Stovepipe* culture.
- Culture of rejecting good ideas that originate in manufacturing environment "because we are not manufacturers."

SUGGESTED READING:
- D. Levine, *Reinventing the Workplace: How Business and Employees Can Both Win*, Brookings Institution Press, 1995.
- J.K. Liker, *The Toyota Way: 14 Management Principles from the World's Greatest Manufacturer*. McGraw-Hill Professional, 2004.
- T. Mikami, *Management and Productvity Improvement in Japan*, Tokyo: JMA Consultants Inc., in Co-operation with Japan Management Association, 1982.
- J.M. Morgan and J.K. Liker, *The Toyota Product Development System*. New York: Productivity Press, 2006.

LEAN PRINCIPLE 6: **Respect for People**
ENABLER: **2. Build an Organization Based on Respect for People**.
SUBENABLER(S) AND USE RANKINGS: **11. Do not manage from cubicle; go to the spot and see for yourself**.   **(U = 0.17)**
VALUE PROMOTED: More effective management, quicker and more effective identification of problems, ability to fix problems more quickly with less waste.
WASTE PREVENTED: All categories of waste.
EXPLANATION: Every manager faces the following choice: to manage from his or her office, demanding that all information about a problem (and solution) be brought to his or her attention and decision; or to go to the site of the problem, see it firsthand, identify it, possibly mentor root cause analysis, ask questions, and help eliminate the problem once and for all. This subenabler promotes the latter practice, which is called *gemba* (Japanese), meaning "go and see." "The first step of any problem-solving process, development of a new product, or evaluation of an associate's performance is grasping the actual situation, which requires 'going to *gemba*'" [Liker, 2004, p. 224]. The short word means more than just "seeing first hand." It also means becoming involved. In order to motivate managers and employers to get to the root cause of any problem, Tadashi Yamashina, president of the Toyota Technical Center, suggests: "[*Gemba*] is more than going and seeing. [It also means having to ask questions:] 'What happened? What did you see? What are the issues? What are the problems?' Within the Toyota organization in North America, we are still just going and seeing. 'OK, I went and saw it and now I have a feeling.' But have you really analyzed it? Do you really understand what the issues are? At the root of all of that, we try to make decisions based on factual information, not based on theory. Statistics and numbers contribute to the facts, but it is more than that. Sometimes we get accused of spending too much time doing all the analysis of that. Some will say 'Common sense will tell you. I know what the problem is.' But collecting data and analysis will tell you if your common sense is right." [Liker, 2004, p. 225]
SUGGESTED IMPLEMENTATION: This subenabler is relatively easy to implement with the right enterprise leadership, training of managers, and incentivizing the right behavior among managers.

LAGGING FACTORS:
- Enterprise leadership not aware of the practices and benefits of *gemba*.
- Traditional management style from a cubicle.

SUGGESTED READING:
- J.K. Liker, *The Toyota Way: 14 Management Principles from the World's Greatest Manufacturer*. McGraw-Hill Professional, 2004.
- J.M. Morgan and J.K. Liker, *The Toyota Product Development System*. New York: Productivity Press, 2006.

LEAN PRINCIPLE 6: Respect for People

ENABLER: **2. Build an Organization Based on Respect for People**.

SUBENABLER(S) AND USE RANKINGS:
13. Build a culture of mutual support (there is no shame in asking for help). (U = 0.36)

VALUE PROMOTED: Faster flow of value, less waste, improved collegiality and teamwork.

WASTE PREVENTED: Primarily waiting, and to some degree all other categories of waste.

EXPLANATION: High-powered organizations place a high value on mutual support. Mutual support should work in every direction: engineers supporting their colleagues; supervisors and managers supporting employees, other managers, and superiors; buyer's employees supporting suppliers; all employees supporting the customer; etc.

In large PD organizations, both the formal and *tribal knowledge* is deep and broad, and usually there are numerous senior employees who have answers to many questions at their fingertips. Many problems repeat themselves, and many have been solved or addressed in the past. Keeping this knowledge hidden is detrimental to an organization's health, wasteful of precious program time, and conducive to the waste of *reinventing the wheel*. In contrast, sharing knowledge freely with colleagues benefits everybody: It promotes collegiality and teamwork, saves the value creation time, improves team performance, and assists junior employees to come "up to speed" faster. Most people enjoy being asked for advice and sharing knowledge. Employees should be informed that "sitting on a question" without seeking an immediate answer is bad, and that there is no shame in asking for help. This is particularly important for new hires.

SUGGESTED IMPLEMENTATION: This subenabler is relatively easy to implement: It should be enough to send a message from the middle managers to the effect: "If you have a question that you alone cannot answer in a reasonable time, immediately seek the person who will help you." Another message to all employees should be that "everybody is expected to answer questions from any colleague in a reasonably short time, fully, honestly, openly, in a friendly way, without blaming the person asking." This must be implemented reasonably: A super busy employee working on a critical task should not be interrupted and expected to drop everything to answer questions. But the opposite should also be true: Any simple question requiring a simple answer should be supportively addressed without undue delay.

A concern of some managers that the system may be abused and employees will start asking questions instead of thinking for themselves is rarely justified. In the SE environment, most employees are professionals. A rare abuse is easy to correct through a talk with a manager, and, in a truly exceptional case, by shifting the employee to a less challenging position. Organizations should be designed to benefit the vast majority of good employees rather than to police the tiny minority of poor performers.

LAGGING FACTORS:
- The culture of rivalry, keeping all information to oneself.
- Fear of layoffs.
- Adversarial human relations.

SUGGESTED READING:
- J.M. Morgan and J.K. Liker, *The Toyota Product Development System*. New York: Productivity Press, 2006.
- R. Leopold, *The Iridium Story: An Engineer's Eclectic Journey*, Minta Martin Lecture, MIT Department of Aeronautics and Astronautics, April 23, 2004.

LEAN PRINCIPLE 6: Respect for People
ENABLER: 2. Build an Organization Based on Respect for People.
SUBENABLER(S) AND USE RANKINGS: 14. Prefer physical team co-location to the virtual co-location. (U = 0.44)
VALUE PROMOTED: Vastly better communication, coordination, planning, and resolution of issues in real time
WASTE PREVENTED: All categories of waste.
EXPLANATION: In large complex PD programs, the need to plan, discuss, brainstorm, negotiate, coordinate, and resolve issues and risks occurs frequently indeed. The subenabler promotes face-to-face meetings of co-located stakeholders, where possible. Such meetings are incomparably more effective than even the best video conferencing technology. There is a certain intangible psychological magic in face-to-face interactions, which is conducive to more effective decision making. Human beings have practiced such interactions for tens of thousands of years and developed them to a high level of art. Body language is a critical component of communication, especially in difficult and stressful situations (which are frequent in PD work). People meeting in one room engage both sides of the brain, which is said to improve creativity. This is also important for brainstorming and negotiations. It engages the power of teamwork and consensus building. Cross-talk or even cross-looks among individuals in the room are all important, if intangible, signals of communications. In contrast, virtual meetings are not nearly as effective. Most people speaking to a camera tend to act in an artificial manner, "playing the part." People tend to be more bureaucratic and more formal, and thus, less creative. Usually only the speaker's face is seen, and the reactions of the colleagues are not. The quality of a meeting goes down as bandwidth narrows. Telecoms are less effective than video conferencing and electronic chats less than telecoms, and so on. Of course, co-location alone is not a guarantee of a good meeting, but the lack of co-location is always a detriment. Looking at the history of large defense programs, the best performing ones were indeed the programs executed in one physical location (e.g., Skunk Works in Palmdale; Northrop aircraft in Hawthorne, Hughes Space and Communications in El Segundo, TRW in Redondo Beach, and many others). The recent unfortunate trend of virtual meetings was made possible by modern communication electronics and was promoted by globalization and mergers of companies into giants operating *in all 50 states*. Some companies have managed to resist the negative trends, e.g., Rockwell Collins and Thales (see Chapter 8).

SUGGESTED IMPLEMENTATION: For important meetings, perform a quick cost-benefit analysis of co-location versus virtual meeting. Evaluate the cost of transporting key people to a common least expensive location (direct travel cost and the cost of time during travel, hotel, per diem) and compare it to the cost of potentially less-than-perfect decisions made in the absence of co-location. While the latter may be difficult to estimate accurately, past experience should be used as a guide: Just think of all those PD programs that failed, or experienced significant cost and schedule overruns, because of bad decisions, lack of consensus, lack of good planning, failure to address and resolve issues fully, bad communications, and many other issues. The Lean approach strongly promotes co-location for all planning and *integrative events*.

LAGGING FACTORS:
- Large geographical distribution of stakeholders.
- False belief that electronic and automated means of communication are as good as direct human communications.
- High cost of travel.
- Objective obstacles to travel (e.g., natural disasters and airline strikes).

SUGGESTED READING:
- J.M. Morgan and J.K. Liker, *The Toyota Product Development System*. New York: Productivity Press, 2006.
- E.M. Murman, *Lean Aerospace Engineering*, Littlewood Lecture AIAA-2008-4, January 2008.
- D. Secor, *Implementation of Lean Enablers*, LAI Plenary, Dana Point, March 23, 2010.
- B.R. Rich and L. Janos, *Skunk Works*. Little & Brown, 1994.

LEAN PRINCIPLE 6: Respect for People

ENABLER: **3. Expect and Support Engineers to Strive for Technical Excellence**.

SUBENABLER(S) AND USE RANKINGS:
1. Establish and support Communities of Practice. (U = 0.67)

VALUE PROMOTED: Effective sharing of knowledge, experience, and wisdom among engineers of like specialty and interests promotes value creation.

WASTE PREVENTED: All categories of waste.

EXPLANATION: The high complexity of modern PD programs and the vast rate of change of engineering and scientific knowledge require that individual engineers and managers should have accessible opportunities to draw from and share their knowledge, experience, and wisdom with their peers. The most effective platform for informal sharing is a community of practice promoted by this subenabler. It can be organized at a number of levels: within a department among engineers of the same specialty, within an enterprise, within a local chapter of a professional society, and within national or even international professional societies, as listed below. Engineers participating in such communities of practice tend to stay *au courant* (up to date) on latest developments in their profession, department, or enterprise, learn faster, and overall become better engineers than those who do not.

SUGGESTED IMPLEMENTATION:
- *Department level:* The engineers of a like specialty meet periodically to discuss their case studies, good ideas as well as failed ones, share solutions or actions, dos and don'ts, and their experiences and wisdom. These meetings should be informal, pleasant, and free of work hierarchy, while focused on honest sharing of knowledge, experiences, and wisdom. Snacks or meals are always welcome and are a good investment. Better case studies ought to be written up and made available in a database to all engineers of a given specialty.
- *Enterprise level:* Just like the departments, but involving all engineers of a given specialty from all divisions of an enterprise. Normally, this community of practice needs a bit more formality than the departmental level, e.g., emailing list, regular meeting times, a convenient location, and advertised agenda.
- *Local chapter of a professional society:* Professional societies have a long and rich tradition of organizing local chapters for sharing knowledge. These chapters organize periodic meetings as occasions for both networking and sharing knowledge via seminars, tutorials, and workshops. The knowledge shared tends to be less company-specific and more generic.

- *National/international professional society:* Engineering professional societies operating at national or international levels offer numerous benefits for a community of practice: conferences, tutorials, workshops, short courses, journals, books and newsletters, peer review, lobbying services, and even financial services (credit unions and insurances).

LAGGING FACTORS:
- Departments that are so busy with daily work that no time is left over for any other activity.
- Remote geographical location, making access to a common site impractical.

SUGGESTED READING:
- D. Lempia, *Using Lean Principles and MBE in Design and Development of Avionics Equipment at Rockwell Collins*, *26ᵗʰ International Congress of the Aeronautical Sciences,* Anchorage, AK, September 16, 2008, paper 6.7.3.
- J.M. Morgan and J.K. Liker, *The Toyota Product Development System*. New York: Productivity Press, 2006.

LEAN PRINCIPLE 6: Respect for People

ENABLER: 3. Expect and Support Engineers to Strive for Technical Excellence.

SUBENABLER(S) AND USE RANKINGS:
2. Invest in Workforce Development. (U = 0.83)

VALUE PROMOTED: Well-developed engineers are capable of creating value more competitively.

WASTE PREVENTED: Well-developed engineers are capable of reducing all kinds of waste.

EXPLANATION: The engineering profession faces two continuous challenges: exponentially growing engineering knowledge and the growing complexity of products and PD programs. Both require that every engineer should continuously develop his or her knowledge and skills. In order to stay competitive, PD companies should provide well-designed opportunities for development and treat associated costs as an investment. Popular development activities include formal training, mentoring, rotations through key functions, formal university-level courses, certificate and degree programs, short courses and tutorials, webinar courses, professional society lectures, participation in communities of practice, and, in exceptional cases, scholarly sabbaticals and internships at scholarly institutions.

"In a lean system, people learn best from a combination of direct experience and mentoring. Excellent engineers that fit in with a high-performance PD do not graduate from college ready baked to handle important projects; they are built slowly from scratch. Toyota has always recognized this truth and has developed rigorous selection and training processes to support it." [Morgan and Liker, 2006, p. 163].

SUGGESTED IMPLEMENTATION: Each corporation should develop a formal program and budget for workforce development. The educational path should be well designed, consistent with a company's long-term strategy and mission. Large companies can enter into agreements with teaching institutions to offer well-designed qualified education at a discount.

LAGGING FACTORS:
- Expectation that engineering education (college level or MS level) is sufficient to immediately start productive engineering work.
- No workforce development opportunities in the workplace.
- Employees driven too hard to have any time or energy left for development.
- Only on-the-job training without mentoring (which really means no training, and junior engineers repeating the same mistakes all the time, causing program delays).

SUGGESTED READING:

- J.M. Morgan and J.K. Liker, *The Toyota Product Development System*. New York: Productivity Press, 2006.

- D. Lempia, *Using Lean Principles and MBE in Design and Development of Avionics Equipment at Rockwell Collins*, 26th *International Congress of the Aeronautical Sciences,* Anchorage, AK, September 16, 2008, paper 6.7.3.

- D. Secor, *Implementation of Lean Enablers*, LAI Plenary, Dana Point, March 23, 2010.

LEAN PRINCIPLE 6: Respect for People

ENABLER: 3. Expect and Support Engineers to Strive for Technical Excellence.

SUBENABLER(S) AND USE RANKINGS:
3. Assure tailored Lean training for all employees. (U = 0.21)
4. Give leaders at all levels in-depth Lean training. (U = 0.13)

VALUE PROMOTED: Faster implementation to the dramatic benefits of Lean Thinking in SE and entire PD.

WASTE PREVENTED: All categories of waste.

EXPLANATION: While the understanding of Lean Thinking in production and manufacturing is now widespread, Lean in Product Development and specifically in Systems Engineering is not yet popular either in industry or government, even though the literature on these subjects is already significant, and cases of success are well documented. If PD enterprises wish to remain competitive, they should acquire Lean knowledge as soon as possible using the most effective means. This statement applies equally well to government and industry. Training by experts (initially external, until in-house experts are well developed) is probably the most effective way to acquire the first level of knowledge (understanding of fundamentals). Subenabler 6.3.3 promotes tailored lean training for all employees, and subenabler 6.3.4 for leaders. The training scope for these groups is listed below. Liker [2004, pp. 182–183] strongly promotes a special training for leaders: "The leader's goal at Toyota is to develop people so they are strong contributors who can think and follow the Toyota Way at all levels in the organization. The leader's real challenge is having the long-term vision of knowing what to do, the knowledge of how to do it, and the ability to develop people so they can understand and do their job excellently. The payoff for this dedication is more profound and lasting to a company's competitiveness and longevity than using a leader merely to solve immediate financial problems, make the correct decision for a given situation, or provide new short-term solutions to bail a company out of a bad situation. A company growing its own leaders and defining the ultimate role of leadership as 'building a learning organization' lays the groundwork for genuine long-term success."

SUGGESTED IMPLEMENTATION: All PD stakeholders should be organized into groups of different levels and needs for tailored Lean training. Leaders should form a separate group so that they can ask questions without losing face. All training courses for all levels should include the following basics (it is assumed that Lean manufacturing knowledge is already in place; if not, this should be acquired first): the concepts of value in PD, PD waste (including examples of waste), the process of creating PD value without waste captured into the six Lean

Principles, fundamentals of Value-Stream Mapping in PD, definition of Lean SE, a quick review of the Lean Enablers for SE, a review of the benefits from Lean PD, and a few case studies from published literature. Experience demonstrates that this all can easily be covered in a one-day tutorial. After this tutorial, leaders should be sufficiently convinced and excited to support subsequent Lean implementation in their enterprises, and others should be sufficiently excited to Lean to seek more knowledge. Then, additional sessions should be conducted in two- to four-hour advanced modules devoted to specific aspect of Lean PD/SE. Not all modules are needed by all managers and engineers—the modules should be tailored to different audiences as needed. The total number of modules practiced by the author is nine, as follows (the actual breakdown of the topics and delivery should be left up to the individual instructors):

- Capturing customer need and requirements, including *unspoken requirements*, and Lean development of requirements and other aspects of value.
- Planning and Frontloading PD.
- Lean SE and Lean Enablers for SE.
- Lean final engineering (engineering of parts and assembly), including Lean *Design for Manufacturing, Assembly and Testing (DFMAT)*
- Lean integration, verification and testing, and validation.
- Lean Quality Assurance.
- Lean Supply Chain.
- Lean Accounting and Metrics.
- Lean Administration (a.k.a. Lean Office).

LAGGING FACTORS:
- No plans for Lean training.
- Too busy *fighting fires* to implement Lean (this is when it is most critical to apply Lean Thinking).
- A viewpoint that engineers are capable of learning only a few simple Lean rules and must not be overwhelmed with too many details of Lean.
- Leaders not willing to look at anything other than short-term financial performance.

SUGGESTED READING:
- J.P. Womack and D.T. Jones, *Lean Thinking*. Simon & Shuster, 1996.
- D. Lempia, *Using Lean Principles and MBE in Design and Development of Avionics Equipment at Rockwell Collins*, 26th *International Congress of the Aeronautical Sciences*, Anchorage, AK, September 16, 2008, paper 6.7.3.
- J.M. Morgan and J.K. Liker, *The Toyota Product Development System*. New York: Productivity Press, 2006.
- O. Terrien, *An Experience Report at Thales Aerospace, France "The Lean Journey"*, Thales Aerospace, INCOSE 2010 IS, Chicago.

LEAN PRINCIPLE 6: Respect for People

ENABLER: 4. Nurture a Learning Environment.

SUBENABLER(S) AND USE RANKINGS:
1. Perpetuate technical excellence through mentoring, training, continuing education, and other means. (U = 0.82)
2. Promote and reward continuous learning through education and experiential learning. (U = 0.36)
4. Pursue the most powerful competitive weapon: the ability to learn rapidly and continuously improve. (U = 0.55)

VALUE PROMOTED: Sustain competitiveness by rapid effective learning and improving in product development faster than competitors.

WASTE PREVENTED: All categories of waste.

EXPLANATION: As described under subenabler 6.3.2, investment in workforce development is one of the best long-term investments companies can make if they desire to be competitive. This is particularly important for organizations involved in high-tech PD, where knowledge explodes exponentially and rapid learning is critical for success. For such companies effective continuous learning and improvement is not an option; it is a critical aspect of a Lean organization. The present three subenablers (6.4.1, 6.4.2, and 6.4.4) address three mutually complementary aspects of learning for technical excellence. Subenabler 6.4.1 is generic: It promotes an enterprise-wide implementation of best learning practices, including mentoring, training, continuing education, and other means. Subenabler 6.4.2 is more specific: It calls for an explicit practice of rewarding continuous learning through formal education and experience-based learning. Finally, enabler 6.4.4 calls for creating and sustaining a company environment, infrastructure, and culture that permits rapid learning.

SUGGESTED IMPLEMENTATION: The companies involved in PD should implement the entire menu of learning activities, including:
- Formal in-house training and mentoring by more experienced engineers.
- Rotations of engineers through key functions.
- Formal university-level courses.
- Advanced certificate and degree programs.
- Lectures, short courses and tutorials offered by professional societies and experts.
- Participation in communities of practice.
- For exceptional individuals, scholarly sabbaticals and internships at scholarly institutions.

A special budget and infrastructure are needed to administer these educational activities. A system of rewards should be implemented, such as reasonable time off for education, easy access to courses and instructors, an easy-to-use rotation system, bonuses for tangible learning outcomes, and an information system (on company intranet) about local educational events such as Ed fairs.

When a situation calls for rapid learning and improvement (e.g., when a new technology, tool, or competition appears), the company should be prepared to quickly organize a short course, a lecture by an expert, or similar educational experience, and invite stakeholders (employees and possibly key suppliers) to attend.

LAGGING FACTORS:

- Not enough budget for continuous learning.
- Employees driven so hard that no time or energy is left for education.
- Continuous learning not rewarded or appreciated.
- Lack of education infrastructure.
- Lack of policies and recognition for rotations and mentoring.
- Companies ignorant about the local university educational programs.

SUGGESTED READING:

- J.M. Morgan and J.K. Liker, *The Toyota Product Development System*. New York: Productivity Press, 2006.

LEAN PRINCIPLE 6: Respect for People
ENABLER: 4. Nurture a Learning Environment.
SUBENABLER(S) AND USE RANKINGS: 3. Provide knowledge experts as resources and for mentoring. (U = 0.45)
VALUE PROMOTED: Wisdom and experience of experts promotes excellence throughout the PD enterprise.
WASTE PREVENTED: All categories of waste.
EXPLANATION: Practically all large PD enterprises have numerous in-house experts. Normally, these are senior people of recognized wisdom and achievement. The present subenabler promotes making these experts available for internal consultations and for mentoring more junior engineers and managers, just as university professors hold office hours for advising students. This practice is conducive to sharing of the wisdom of experts across the enterprise. Liker [2004, p. 182] describes an interesting practice at Toyota: "Toyota leaders, by having a combination of in-depth understanding of the work and the ability to develop, mentor, and lead people, are respected for their leadership abilities. Toyota leaders seldom give orders. In fact, the leaders often lead and mentor through questioning. The leader will ask questions about the situation and the person's strategy for action, but they [sic] will not give answers to these questions even though they have [sic] the knowledge. . . . The roots of Toyota leadership go back to the Toyoda family who developed Toyota Way Principle 9: 'Grow leaders who thoroughly understand the work, live the philosophy, and teach it to others.'"
SUGGESTED IMPLEMENTATION: • Prepare a directory of in-house subject matter experts listing their fields and access (names, phone, email, physical location, office hours for consultation). • Verify that the experts are willing to devote a few hours per week to inside consultations. • Make the directory available to junior engineers and managers and encourage them to seek the expert's help. • If practical, employ recently retired experts for this purpose. These retired people often appreciate the recognition and also enjoy passing their wisdom on to the following generation. The enterprise may benefit from their experience and wisdom, and the cost of a few hours per week is minimal. • Let new engineers capture the wisdom of "gray hair" experts. Use a knowledge management strategy and do not wait until the experts retire or leave the company.

LAGGING FACTORS:

- Experts overutilized in current active programs with no time left for sharing.
- Experts not willing to share.
- Experts saying: "when I was young I did not ask for help, so why should I now help others?"
- Fear of layoff: "If I tell them everything I know, they will let me go."

SUGGESTED READING:

- J.K. Liker, *The Toyota Way: 14 Management Principles from the World's Greatest Manufacturer*. McGraw-Hill Professional, 2004.

LEAN PRINCIPLE 6: Respect for People
ENABLER: 4. Nurture a Learning Environment.
SUBENABLER(S) AND USE RANKINGS: **5. Value people for the skills they contribute to the program with mutual respect and appreciation. (U = 0.45)**
VALUE PROMOTED: Trust and respect that every employee will do his or her job so that we are successful as a company.
WASTE PREVENTED: All categories of waste.
EXPLANATION: Sam Heltman, Senior Vice President of Administration at Toyota Motor Manufacturing, North America (one of the first five Americans hired by Toyota, Georgetown), said it well: "Respect for people and constant challenging to do better—are these contradictory? Respect for people means respect for the mind and capability. You do not expect them to waste their time. You respect the capability of the people. Americans think teamwork is about 'you liking me and I liking you'. Mutual respect and trust means I trust and respect that you will do your job so that we are successful as a company. It does not mean we just love each other." [Liker, 2004, p. 184] Deming [1982, p. 118] addressed the case of a "lone worker," also relevant here: "There are abundant examples of people that cannot work well in a team, but who demonstrate unquestionable achievement in the form of respect of colleagues and of peers, through inventions and publications in scientific journals. Such a man may make fabulous contributions to the company as well as to knowledge. The company must recognize the contributions of such people, and provide assistance to them."
SUGGESTED IMPLEMENTATION: Without exception, all people at all levels like to be appreciated and respected. Respect, appreciation, and recognition are always conducive to extra effort and better work morale. Individuals with particularly significant contributions should receive particularly significant appreciation. Appreciation should be indicated right away and not only during annual evaluations. All employees should receive respect for their normal work. Implementation of this subenabler should be enterprise-wide, promoted by all levels of management.

LAGGING FACTORS:
- Position that "the job and pay alone are sufficient signs of appreciation."
- Bureaucratic culture in which the lack of trust, honesty, and respect is compensated with demands for bureaucratic evidence.

SUGGESTED READING:
- J.K. Liker, *The Toyota Way: 14 Management Principles from the World's Greatest Manufacturer*. McGraw-Hill Professional, 2004.
- W.E. Deming, *Out of the Crisis*. Massachusetts Institute of Technology, Center for Advanced Engineering Study, 1982.
- J.M. Morgan and J.K. Liker, *The Toyota Product Development System*. New York: Productivity Press, 2006.

LEAN PRINCIPLE 6: Respect for People
ENABLER: **4. Nurture a Learning Environment**.
SUBENABLER(S) AND USE RANKINGS: **6. Capture learning to stabilize the program when people transfer elsewhere or leave. (U = 0.09)**
VALUE PROMOTED: Easy transition of responsibility and knowledge from one individual to his or her successor, with minimum waiting and without the waste of rework or *reinventing the wheel*.
WASTE PREVENTED: Rework and waiting.
EXPLANATION: Some PD programs last 10–15 or more years. A notorious problem in such programs is keeping the program knowledge stable when people move on, retire, become ill, or pass away. Unless the enterprise has captured the learning and knowledge well, any change in a critical stakeholder is likely to introduce instability into the program: Colleagues discover that an individual was the only one who knew, and now, after the person left, nobody seems to know what to do. Knowledge must be reconstructed, causing rework and waiting waste. This may occur even in shorter programs when unfortunate events (deaths, accidents, career moves) cause staff instability. This subenabler calls for a policy of capturing enough data and learning to keep the program stable when people transfer elsewhere or leave. Ideally, knowledge, information, and learning should be well recorded so that a new person can seamlessly take over when the predecessor leaves for whatever reasons.
SUGGESTED IMPLEMENTATION: • Have a formal policy for capturing learning and knowledge and do not rely on a lone individual, which can create a single point of failure. • For key people, institute the system of a deputy or associate who is mentored by a key person, captures his or her knowledge and learning, and is ready to step into the necessary position at a short notice. • Demand checklists and records and excellent workplace organization (often called 5 Ss after the Japanese practice) among the leaders. Balance the quality of these documents against excessive bureaucracy. Keep enough records for another person to be able to take over on a short notice. • Develop and communicate knowledge management strategy at both enterprise and program levels. • Some larger U.S. government PD contracts explicitly call for such practices.

LAGGING FACTORS:

- Lack of policy of capturing and easily transitioning learning and knowledge from one person to his or her successor.
- The culture of everybody keeping knowledge hidden to minimize the risk of layoff.

SUGGESTED READING:

- J.M. Morgan and J.K. Liker, *The Toyota Product Development System*. New York: Productivity Press, 2006.

LEAN PRINCIPLE 6: Respect for People
ENABLER: 4. Nurture a Learning Environment.
SUBENABLER(S) AND USE RANKINGS: 7. **Develop Standards paying attention to human factors, including reading and perception abilities.** ($U = -0.18$) 8. **Immediately organize a quick training in any new standard.** ($U = -0.27$)
VALUE PROMOTED: Effective procedures that are actually read and followed to make processes Lean: consistently and predictably robust, with quality outcomes and minimum waste. Ability to plan and cost effort reliably.
WASTE PREVENTED: All categories of waste.
EXPLANATION: (Also read subenablers 5.6.1–5.6.3) All repeatable processes should be standardized, so that they can be executed predictably, robustly, waste-free, *right the first time*. A process that is unpredictable acts like sabotage on the program planning: It makes it impossible to estimate process effort, cost, and outcome. Regretfully, many standard procedures suffer from poorly written texts, lack of process optimization, and their release takes a long time, usually accompanied by significant bureaucracy. These two subenablers (6.4.7 and 6.4.8) present a good practice for developing and implementing an effective standard procedure.
SUGGESTED IMPLEMENTATION: An effective standard procedure should be developed using the following steps: 1. Optimize the process steps, possibly by trial and error, using the best employee(s) available, and pay attention to human factors. 2. Draft the procedure and test it on the least experienced employee(s). If the least experienced person is able to follow the procedure, it is probably clear enough. Often, that is not the case. Then collect questions and comments from these employees. 3. Redraft the procedure based on the comments and questions identified in (2). 4. Finish the draft, make it clear, and edit to be an attractive document, with visuals and examples of use (e.g., fill out sample forms with data), if applicable. 5. Release the new procedure efficiently and without excessive bureaucracy, remembering that every procedure is to be a living document, updated with new good ideas and lessons learned. If the release is a bureaucratic process taking months or years (during which time better steps are created but are not practiced because they violate the procedure, which is demoralizing), the organization may be better off not having the formal procedure in the first place. Perhaps an informal dissemination might be more effective.

6. Offer a short training to all process stakeholders (often a few minutes during the weekly meeting is enough), saying "we have this new procedure for process X; the reason for issuing it is [...]; here are the critical dos and don'ts of the process; you find the procedure in [location]; the experts in the procedure available to answer your questions are [...]."

7. For critical procedures, offer a test in the procedure, including the following questions:

   a. List the steps, the dos and don'ts.

   b. Describe how you can tell that the outcome is correct.

   c. Describe how you can tell that the outcome is not correct.

   d. What corrective action are you supposed to undertake immediately if the outcome is not correct?

   e. If the corrective action did not help, who do you notify/ask for help?

8. Continue monitoring the procedure use and outcomes and mentor the employees.

The need for attractive, easy-to-follow text with good visuals and examples cannot be overemphasized. A lengthy technical text filled with technical descriptions, jargon, and tables of data is difficult to understand and is time-consuming. As Liker [2004, p. 244] points out: "A picture is worth a thousand words. Acting on the fact that people are visually oriented, new employees at Toyota learn to communicate with as few words as possible and with visual aids."

The high negative use rankings of these two subenablers indicate that good procedures are rare.

LAGGING FACTORS:

- Lack of procedures.

- The culture of "everything we do is unique, standardization does not apply to us."

- Procedures created in a rush by remote managers, without being optimized or tested. Such procedures tend to be ignored.

- Lack of training in procedures.

- The culture of telling employees: "Here are the ... volumes with all procedures; you are required to become familiar with all of them and practice them" (the best way to have the procedures ignored).

- Long release cycle for procedures, causing the procedures to become obsolete, static, and ignored.

SUGGESTED READING:

- J.K. Liker, *The Toyota Way: 14 Management Principles from the World's Greatest Manufacturer*. McGraw-Hill Professional, 2004.

- J.M. Morgan and J.K. Liker, *The Toyota Product Development System*. New York: Productivity Press, 2006.

LEAN PRINCIPLE 6: Respect for People

ENABLER: 5. Treat People as Most Valued Assets, not as Commodities. (U = 0.70)

SUBENABLER(S) AND USE RANKINGS: None.

VALUE PROMOTED: Engaged, creative, entrepreneurial, and empowered workforce ready to take on any challenges enthusiastically.

WASTE PREVENTED: Great workforce is capable of eliminating all waste categories efficiently.

EXPLANATION: This may be the most important enabler among the entire set presented in this volume. And it may be the oldest: It was formulated by Edward Deming over 20 years ago, but has been known throughout human history [Deming, 1982]. It appears self-evident, but its use ranking of 0.70 is still less than perfect "2," leaving room for improvement. The enabler should be regarded as an important step in achieving an enthusiastic, creative, entrepreneurial, engaged, and empowered workforce. We want employees who are so involved in their work using both sides of the brain that they tend to come to work on Monday before the official start time because they can't wait to start putting in place the great ideas that have occurred to them over the weekend. These employees who treat work challenges the same way as mountain climbers, ocean sailors, or major league players see their sport challenges: to be overcome by all means as if their life depended on it, because they get great satisfaction from it. This is a rare feeling, but it does happen if work is led by an extraordinary leader who is trusted and loved by the team and who knows how to treat people. The individuals who are lucky enough to work for such a leader know the feeling exactly. The feeling is known among many front line solders. It is the feeling that all team members are voluntarily aligned for a common goal, and all feel indispensible.

Contrast that with an environment that, regrettably, occurs in some large programs: bureaucratic, focused on production of artifacts rather than great engineering, lasting so many years so that nobody has any sense of making progress, led by characterless bureaucrats who all sound alike, surrounded by a rich menu of waste, and unable to revolt against it, ... (add your own frustrations).

SUGGESTED IMPLEMENTATION:
- In promotion, place big value on leadership and human skills.
- Treasure and reward good leadership rather than bureaucracy.
- Respect all people.
- Never hire people as if they were commodities.
- Groom, rotate, and support the best leaders (those trusted by their peers).
- Study the literature on leadership.

LAGGING FACTORS:
- Bureaucratic selection and management of people.
- Selecting new hires for expedient need rather than competence, experience, and interpersonal skills.
- Focus on artifact production rather than great engineering of systems.
- Long programs that make progress invisible.
- Dissolved management.
- Focus on profits rather than technical work.
- Complacency with waste, or worse: incentives for waste.

SUGGESTED READING:
- T. Rockwell, *The Rickover Effect: The Inside Story of How Adm. Hyman Rickover Built the Nuclear Navy*. John Wiley & Sons, 1995.
- J.K. Liker, *The Toyota Way: 14 Management Principles from the World's Greatest Manufacturer*. McGraw-Hill Professional, 2004.
- J.M. Morgan and J.K. Liker, *The Toyota Product Development System*. New York: Productivity Press, 2006.
- E.M. Murman, *Lean Aerospace Engineering*, Littlewood Lecture AIAA-2008-4, January 2008.
- W.E. Deming, *Out of the Crisis*. Massachusetts Institute of Technology, Center for Advanced Engineering Study, 1982.

Chapter **8**

General Guidance for Implementation

"I really liked how core team members and stakeholders collaborated to work through questions, assumptions, risks, issues, and dependencies in such a short period of time [which] typically can take days, weeks, or months to work through!"
—from a customer participating in a Lean kickoff meeting at Rockwell Collins.

8.1 GENERAL GUIDANCE FOR IMPLEMENTING LEfSE

The large number of subenablers (147) sometimes meets with a humorous reaction that "being this long, it is not Lean." However, the LEfSE cover a large spectrum of practices in complex technology environments, with a general focus to improve program value and stakeholder satisfaction and reduce waste, delays, costs, and frustrations. LEfSE promote the culture of trust, openness, respect, empowerment, teamwork, good communication and coordination, and drive for excellence. They encourage healthy relationships between all stakeholders, better coordination between parties handling any complex transaction, streamlined work flow while promoting robustness and *right the first time*, and best methodologies for complex system design. They place emphasis on good preparations, planning, frontloading, and preventive measures. They emphasize process optimization, standardization, continuous improvement, and long-term thinking. They describe best practices for human resource management, creation of communities of practice and knowledge

Lean for Systems Engineering with Lean Enablers for Systems Engineering,
First Edition. Bohdan W. Oppenheim.
© 2011 John Wiley & Sons, Inc. Published 2011 by John Wiley & Sons, Inc.

databases, capture of lessons learned, and continuous education and learning. That is a large trade space. It is not surprising that a set of useful actionable enablers needed to cover all these SE aspects must be comprehensive indeed. A much shorter version would have the value of slogans and not be actionable enough for practical implementation.

As much as the LEfSE represent a significant addition to the traditional SE process, they should not be regarded as the solution to all SE and PD problems or to have the power to eliminate program failures. Of course, much more is needed for that level of success. However, the LEfSE establish a high-bar asymptote of program performance and a good motivation to pursue it.

In Chapter 8.2 we quote some early case studies from most dynamic companies. Arguably fragmentary and sometimes lacking scholarly rigor, they do provide testimony for the significant potential of Lean SE and LEfSE to improve PD programs. And these companies should be commended for their early successes.

Lean is a powerful weapon in waste reduction, and the amount of waste in some larger recent programs can be measured in billions of dollars. In this context, a few hours spent by an enterprise team on familiarization with LEfSE may lead to significant savings.

The level of exposure to LEfSE should be left up to enterprise management. A minimum recommended exposure is a four-hour tutorial (or equivalent); it is proven effective for providing participants with a basic familiarity with LEfSE. Alternatively, a self-taught approach may be adopted, e.g., reading this book or watching the media of choice available on the Lean SE WG website (see Appendix 1). This should be followed by self-study, practice, leadership, mentoring, and lots of good examples.

Some subenablers are more actionable than others, and some are easier to implement than others. While different subenablers are intended for different groups of enterprise stakeholders, it is highly recommended that all stakeholders become familiar with all subenablers. It does not take much time to read all of them. Stakeholder awareness of even those least actionable and most difficult to implement should improve work quality and flow, facilitate better work habits, and increase cooperation among all stakeholders. All subenablers are regarded as important to program success, but not equally so for all programs or all enterprises. Some teams are stronger, e.g., in the value capture phase, but weaker in the program execution, and others are vice versa, and so on. Some enablers can be easily implemented by individual engineers, some by small *Kaizen* teams of local stakeholders, and others require management action. Therefore, the selection and priority of the subenablers to implement should be handled individually by enterprise stakeholders. In general, implementation of particular subenablers should be voluntary, pulled by the need to make SE and the program more efficient, in the areas that provide the largest ROI ("biggest bang for your buck").

The case studies described in Section 8.2, in spite of their fragmentary presentation, indicate that implementation of only a small group of subenablers can produce significant savings. Thus, no enterprise should feel pressure to implement "all or nothing." One should look first for the biggest impediments to value capture and creation, or biggest frustrations, and then apply these subenablers that offer the most effective fixes. In tutorials, the author recommends to audiences to read all enablers and subenablers and focus on those that most resonate with the reader.

Many enablers apply to individual employees. These can be implemented immediately on one's own, without asking for any permission.

Appendix 2 contains a mapping of the LEfSE subenablers onto 26 Systems Engineering practices listed in [INCOSE, 2010]. The readers who are more comfortable with these processes rather than with six Lean Principles may find it useful to read the mapped subenablers in Appendix 2 when looking for implementation ideas, and then go back to Chapter 7 to read the detailed explanations.

Communities of practice, such as the INCOSE Lean Systems Engineering Working Group, are available to share ideas and experiences and learn from others. Experts are available in the Working Group and at selected universities for dedicated support.

8.2 EARLY CASE STUDIES

As mentioned earlier, studies of the effectiveness of Lean Systems Engineering should ideally be based on comparisons of program performance with and without Lean, and analyzed with statistics. Measurements are very difficult to obtain for a number of reasons: To-date implementations of Lean are fragmentary; most programs practicing formal Systems Engineering are classified and hard to access by outsiders; most programs involve a large number of stakeholders and organizations lacking a common set of useful metrics; most major programs extend over several years' (some exceeding 10 – 15 years); and where some data exists, its quality and resolution are usually inadequate for our purpose.

In spite of these barriers, some early case studies are available and discussed here. Indeed, all of the results quoted are fragmentary. Some lack scientific rigor. The author includes them because they provide testimony for the significant potential of Lean SE and LEfSE to improve PD programs. The reader may wish to take these early case studies in the same spirit as, arguably, parental advice to study well to get good grades: Common sense indicates that the advice is sound, even if we lack the ability to guarantee that our particular child will do well after graduation.

Since the release of LEfSE to the public in February 2009, the interest in Lean Systems Engineering has grown internationally, accelerating after the INCOSE Best Product Award in February 2010 and again after the Shingo Award for Research and Publication in September 2010. At the time of this writing, about 40 workshops on LEfSE have been completed worldwide, and the Lean SE Working Group of INCOSE has grown to 200 members. Informal contacts with many members in these communities indicate that they are trying to implement selected enablers, although not much hard data is available. INCOSE branches in several European countries and in Israel participated in the development of LEfSE, and they actively promote LEfSE among their members.

Most of the available case studies come from the INCOSE International Symposium in Chicago, July 10–15, 2010, which included four events devoted to Lean

SE and LEfSE (a panel, two meetings, and a paper session), and Lean SE was mentioned in several other events. The growing implementation trend is likely to accelerate with the release of a future version of INCOSE SE Handbook with LEfSE mapped onto process descriptions (Appendix 2 contains a draft of the mapping) and after questions on Lean SE are added to the INCOSE SE Certification exams (both projects have been initiated by INCOSE.) At this time, however, we are limited to the following case studies:[1]

8.2.1 Rockwell Collins, Incorporated (RC)

Rockwell Collins is a company with 20,000 employees worldwide, including over 4800 engineers, specializing in commercial avionics, cabin electronics, and military airborne and surface solutions for U.S. and international markets. Its 2009 annual sales were $4.47B. Rockwell Collins' CEO is a strong Lean proponent. Several RC engineers were among the most active participants in the LEfSE development effort and in the leadership of Lean Systems Engineering. Lean experts at RC in Cedar Rapids, Iowa, are driving an energetic implementation of LEfSE. This implementation is being accomplished by awareness, training, and specific application of Lean Enablers within project planning disciplines. RC includes links to INCOSE Lean Enablers on their company intranet in all searches under Lean Engineering, Lean Systems Engineering, and Lean Project Planning. When employees look for ideas to implement Lean within the engineering practices, Lean Enablers provide rich content. RC looks at the Lean Enablers from a systems approach, understanding that some of the enablers are key to project success. The reception by employees, company leaders, and customers to implemented enablers has been enthusiastic. Many have commented that LEfSE are "good common sense principles." According to Secor [2010], the leader of LEfSE implementation at the company, the results have been so positive the company decided to rollout LEfSE and offer the associated rigorous project planning at all RC regional locations. No numerical data was offered, but the company provided the following quotes from actual customers who participated in the program planning effort:

- "Awesome!! We need this approach on all program kickoffs. Much was achieved in a day."
- "I really liked how core team members and stakeholders collaborated to work through questions, assumptions, risks, issues, and dependencies in such a short period of time ... [(a day), which] typically can take days, weeks, or months to work through!"
- "The timeliness of information was great ... "
- "Great information sharing ... negotiation concept worked very well."

---

[1]The readers are kindly requested to provide results of their own implementation of LEfSE to the author, preferably with concrete data, so that this knowledge can grow. The author's email can be found on the INCOSE Lean SE WG site: www.incose.org/practice/techactivities/wg/leansewg/

- " . . . I think this type of planning session will eliminate scheduling problems down the road."
- "[I liked] having all the functions involved in the project in the same room. Commitments are made face to face [which] solidifies the relationship."

8.2.2 Thales Aerospace, France

Thales Aerospace, headquartered in France, with 13,000 employees worldwide, specializes in defense and commercial cockpit and cabin electronics and defense equipment. Baujard et al. [2010] reported on LEfSE implementation results in one French facility only. Five following enablers were quoted in the paper as having been implemented:

2.2.2. Have cross-functional stakeholders work together to build the agreed value stream.

2.2.5. Use formal value-stream mapping methods to identify and eliminate SE and PD waste, and to tailor and scale tasks.

2.2.13. Plan to use visual methods wherever possible to communicate schedules, workloads, changes to customer requirements, etc.

3.8.1. Use Lean tools to promote the flow of information and minimize handoffs: small batch size of information, small *takt times*, wide communication bandwidth, standardization, work cells, training.

5.3. Use Lessons Learned from Past Programs for Future Programs

According to the quoted results, Lean workshops organized for 5 – 10 people each have achieved an ROI of five and reduced development cycle time, on average, by 20%. The results were found to be so good that a widespread implementation across all functional organizations is now underway. At the time the quoted paper was written, more than a 100 Lean SE workshops have been completed, and 500 people from executive managers to technicians have been trained. The paper authors emphasize that Lean is about initiating behavioral changes and a culture of continuous improvement.

8.2.3 Rafael, Israel

Rafael is one of Israel's largest defense companies with 5000 employees. It creates products for air, land, naval, and security warfare. Zonnenshain [Lean Panel, 2010] presented a partial implementation of Lean Thinking focused on minimizing the time wasted by designers. His presentation did not list specific enablers. Based on the presentation, the author of the present book deduced that approximately 13 following enablers were informally implemented:

1.2.3. Develop a robust process to capture, develop, and disseminate customer value with extreme clarity.

1.3.1. Everyone involved in the program must have a customer-first spirit.

2.2.1. Develop and execute a clear communication plan that covers entire value stream and stakeholders.

2.2.2. Have cross-functional stakeholders work together to build the agreed value stream.

2.2.9. Synchronize work flow activities using scheduling across functions and even more detailed scheduling within functions.

2.3.1. Plan to utilize cross-functional teams made up of the most experienced and compatible people at the start of the project to look at a broad range of solution sets.

2.2.4. Maximize co-location opportunities for SE and PD planning (SE is a part of PD. In this paragraph, the PD should be understood as denoting all PD activities other than SE, including design, development, manufacturing, integration, testing, etc.).

2.5.1. Select suppliers who are technically and culturally compatible.

2.5.2. Strive to develop seamless partnership between suppliers and the product development team.

2.5.3. Plan to include and manage the major suppliers as a part of your team.

3.3.5. Invite suppliers to make a serious contribution to SE, design, and development as program trusted partners.

3.2.5. *Fail early-fail often* through rapid learning techniques (prototyping, tests, digital preassembly, spiral development, models, and simulation).

6.4.1. Perpetuate technical excellence through mentoring, training, continuing education, and other means.

The presenter estimated an overall effect of this Lean effort as a 40% reduction of the time wasted by designers.

8.2.4 EADS and AFIS

The European Aeronautic Defense and Space Company (EADS) is a pan-European aerospace corporation headquartered in the Netherlands. It produces military and commercial aircraft, including Airbus, communications systems, missiles, space rockets, satellites, and related systems. Its approximately 120,000 employees create sales of about 41 billion Euro. Mauritz [2010] an executive of EADS, listed Lean Systems Engineering as the most important imperative in his presentation of the five imperatives of EADS Systems Engineering.

Association Française d'Ingénierie Système (AFIS) is the French branch of INCOSE. Fanmuy [2010] of AFIS briefly described his plans for a comprehensive implementation of LEfSE within the companies served by AFIS.

8.2.5 Early Results from a Study by E. Honour

Honour [2010] presented early results from his comprehensive quantitative PhD research on the Return on Investment of recent SE programs. Among others, he included two rather profound results, as follows:

- Well-conducted and funded SE activities correlate relatively well with cost and schedule compliance and overall program success, but have little effect on the technical quality of the product system (aiming for minimum requirements compliance rather than optimizing technical quality).
- The programs that emphasized technical management and leadership, including program planning, technical reviews, technical control, team leadership, interdisciplinary coordination, common language and goals, risk management, configuration management, and interface management, performed better than those that did not.

In the opinion of this author, both results support the need for LEfSE: Among the 147 LEfSE, several explicitly address the good practices listed by Honour.

8.2.6 Toyota

It is only proper to end the case studies with a tribute to the company that started it all: Toyota. Among the 147 subenablers listed in LEfSE, a good number have been inspired by Toyota's PD practices [Morgan and Liker, 2006]. Again, rigorous data about the individual effectiveness of these enablers at Toyota is not available, but the extraordinary success of Toyota[2] in Lean PD provides strong circumstantial evidence that the Lean approach is worth implementing.

---

[2]As mentioned in footnote 2 in Chapter 1, we refer to the successes before the recent troublesome period.

Glossary of Abbreviations

a.k.a	also known as
AFIS	Association Française d'Ingénierie Système (French Chapter of INCOSE)
AIAA	American Institution for Aeronautics and Astronautics
AR	Acquisition Reform (of DoD)
ASAP	As soon as possible
CAE	Computer Aided Engineering
CE	Concurrent Engineering
CI	Continuous Improvement
CMMI	Capability Maturity Model Integration
CONOPS	Concept of Operations
DFMAT	Design for Manufacturing, Assembly and Testing
DoD	Department of Defense
DODAF	DoD Architecture Framework
EADS	European Aeronautic Defense and Space Company
EdNet	LAI Educational Network
ELOP	Elbit Systems Electro—Optics, Israel
EM	Enterprise Management
FBC	Faster, Better, Cheaper (a former NASA practice)
GAO	Government Accountability Office
GE	General Electric

Lean for Systems Engineering with Lean Enablers for Systems Engineering,
First Edition. Bohdan W. Oppenheim.
© 2011 John Wiley & Sons, Inc. Published 2011 by John Wiley & Sons, Inc.

HALT/HASS	Highly Accelerated Life Test/Highly Accelerated Stress Screen
IPT	Integrated Product Team
ISO	International Standards Organization
INCOSE	International Council on Systems Engineering
JSF	Joint Strike Fighter (F-35)
LAI	Lean Advancement Initiative, formerly Lean Aerospace Initiative, formerly Lean Aircraft Initiative
LEfSE	Lean Enablers for Systems Engineering
LEM	Lean Enterprise Model
LMU	Loyola Marymount University
LPD	Lean Product Development
LPDF	Lean Program Development Flow
LSE	Lean Systems Engineering
LSE WG	Lean Systems Engineering Working Group (of INCOSE)
MBA	Master of Business Administration
MIT	Massachusetts Institute of Technology
MoD	Ministry of Defense (in United Kingdom)
MRM	Mature Robust Modules
NASA	National Aeronautics and Space Administration
NBC	National Broadcasting Company (TV channel)
NVA	Non-value-added
PATI	(meeting) Purpose, Agenda, Time, Intent
PD	Product Development
PDCA	Plan, Do, Check, Act, improvement cycle
PM	Project Management
RAA	Responsibility, Accountability, Authority
RASI	Responsible, Approving, Supporting, and Informing
RFP	Request for Proposals
RNVA	Required Non-Value Added
ROI	Return on Investment
SBIRS	Space Based Infrared System
SE	Systems Engineering, or Systems Engineer
TBD	To be determined
TQM	Total Quality Management
U.K.(UK)	United Kingdom
U.S. (US)	United States
USAF	United States Air Force
VA	Value Added
WG	Working Group

GLOSSARY OF IDIOMS, COLLOQUIALISMS AND FOREIGN EXPRESSIONS *(bold and italicized in the text)*

À priori	(Fr.) knowledge prior to experience
All together	An urgent activity in which all employees participate equally, regardless of their position.
As you fly, test as you fly	Testing under as realistic conditions as possible.
Au courant	(Fr.) up to date.
Batch and queue	The term refers to a traditional production per quota, producing massive batches (inventories) of parts regardless of the immediate need by the next station, with batches waiting in queues.
Big bang approach	Using all required (or massive) resources.
Big guns	Important managers or stakeholders.
Big picture	Seeing or understanding the entire project rather than just own work.
Chicken verus egg	Refers to an engineering iterative method of starting with an arbitrary (or near-arbitrary) point in the tradespace (e.g. assuming the geometry of a part) in order to analyze it, compare the analyzed behavior to performance limits, and keep iterating until all limits are satisfied.
Cost plus	The type of contract where the contractor is reimbursed for all costs (performed labor, materials, etc.), and is paid a percent of the costs as profit. This is in contrast to "fixed price" contracting.
Cycle time	see throughput time
Earned Value	The method of book keeping whereby the contractor is paid for each predetermined deliverable, such as a drawing, or part specifications.
Fail early-fail often	An iteration method based on trial and error, recommended in early design phases, of quickly evaluating a large number of possibilities in the trade space, or quickly testing simple prototypes, when costs are still small.
Five whys	Asking five (or so) times in a row the question "why" trying to discover the root cause of a problem. After each question, an answer reveals the next layer of understanding. It is said to require five questions to identify the root cause in typical modern environments.
Flowdown (flow-down, flow down)	Passing responsibility, an obligation, or action—onto lower level management, subcontract, or employee.

Gemba	(Jap.) meaning "go to the place of action and see for yourself".
Hands off attitude	An attitude in business relations of not communicating with the business stakeholder.
Homework	Refers to the work on tasks in home departments, between the integrative events.
Integrative event	A meeting with well defined agenda to resolve all issues that have appeared during the homework period.
In all 50 states	Refers to contracts whereby the contractor distributes work among facilities in all (or almost all) 50 U.S. states, in order to gain support of as many Congress persons as possible for the program.
Ju-jitsu	(Jap.) The name of popular martial art. Used in the Six Sigma terminology to denote different levels of training.
Kaizen	(Jap.) A small team of local stakeholders tasked to perform a rapid focused improvement of a problem
Kanban	Literally meaning "signboard" or "billboard", is a scheduling signal informing what to produce, when to produce it, and how much to produce.
Know how	Practical knowledge of how to get something done.
Latest, greatest, and gold-plated	A sarcastic expression summarizing some government contracts that are the opposite of frugal.
Low balling	Under estimating.
Monument	A machine, tool or or software that is too big, too expensive, too complex for the needed application and for the flow of work.
One off	Unique.
Over the wall	The practice of sending specifications to a supplier or stakeholder with no follow up, explanations, or clarifications.
Project owner	The person fully responsible for the project.
Reinventing the wheel	The activity of discovering what already has been discovered.
Requirements creep	Uncontrollable growth of requirements.
Right the first time	An execution of a task which produces correct outcomes on the first attempt, without any need for adjustments, corrections, rework, etc.
Sacred	Not to be challenged under any circumstances.
Single piece flow	The flow of work pieces one at a time through all work stations, without batches or inventories.

Stove pipe (stove piped)
: Refers to an organization with rigid boundaries between departments, with hierarchical management, with information flowing up (to next manager) and down (to his/her employee) with restricted flow across the boundaries. Characteristic of traditional organizations.

Takt time, takt period
: The rhythm of assembly line, also the amount of time allocated to each workstation to complete the work, also the rate at which products flow to customers. In this book: the rhytm of homework and integrative events (see above) in the LPDF method.

Traffic cop
: A manager who steers and directs work to and from other stakeholders (employees).

Throughput time (cycle time)
: The amount of time needed to complete the work (of a project, or project segment, or task).

Tribal corporate knowledge
: Unwritten knowledge and set of rules rules about corporate customs and behavior.

Unspoken requirements
: The requirements that the customer failed to include either because they appeared self-evident, or were not known at the time.

War room
: Also called project room. A large room which serves as the headquarter for the project, with plans and other information displayed on the room walls. It is the room used for planning and integrative meetings.

References

T. Allen, *Optimization of Organizational Matrix*, Lecture at Lean Aerospace Initiative Annual Symposium, Los Angeles, California, 2002.

M. Baujard, H. Gilles, and O. Terrien, *An Experience Report at Thales Aerospace, France 'The Lean Journey'*. Paper session on ROI and Lean, INCOSE 2010 International Symposium, July 10–15, 2010, Chicago.

C. Bogan, *Benchmarking Best Practices*, McGraw Hill, New York, 1994.

K. Bozdogan, *Roadmap for Building Lean Supplier Networks (Roadmap Tool)*, Center for Technology, Policy and Industrial Development, Massachusetts Institute of Technology, Cambridge, Massachusetts, March 15, 2004.

W. J. Broad, *Debris Spews into Space After Satellites Collide*. New York Times, Feb. 12, 2009, p. A26.

T. R. Browning, *Modeling and Analyzing Cost, Schedule, and Performance in Complex System Product Development*, Doctoral Thesis in Technology, Management and Policy, Massachusetts Institute of Technology, Cambridge, MA, Dec. 1998.

T. R. Browning, *Value Based Product Development: Refocusing Lean*. IEEE EMS Int. Eng Management Conf. (IEMC), Albuquerque, New Mexico, Aug. 13–15, 2000, pp. 168–172.

J. Byrne, *Editorial*. Business Week, June 23, 1997, p. 47.

A. B. Carter, The Under Secretary of Defense, Acquisition, Technology and Logistics. *Memorandum for Acquisition Professionals*, June 28, 2010.

W. L. Carter, *Process Improvement for Administrative Departments—The Key to Achieving Internal Customer Satisfaction*, BookSurge Publishing, 2008.

Lean for Systems Engineering with Lean Enablers for Systems Engineering,
First Edition. Bohdan W. Oppenheim.
© 2011 John Wiley & Sons, Inc. Published 2011 by John Wiley & Sons, Inc.

J. P. Chase, *Value Creation in the Product Development Process*, Masters Thesis in Aeronautics and Astronautics, Massachusetts Institute of Technology, Cambridge, MA, Dec. 2001.

K. B. Clark and T. Fujimoto, *Product Development Performance; Strategy, Organization, and Management in the World Auto Industry*, Harvard Business School Press, 1991.

K. B. Clark and T. Fujimoto, *The power of product integrity*. Harvard Bus Rev, November-December, 1990, 68(6), 107–118.

D. Clausing, *Total Quality Development: A Step-By-Step University Guide to World-Class Concurrent Engineering*, ASME Press, New York, 1994.

Capability Maturity Model Integration (CMMI), V1.3, Software Engineering Institute, Carnegie Mellon University, October 28, 2010.

R. B. Costello, Office of the Under Secretary of Defense (Acquisition), *Bolstering Defense Industrial Competitiveness: Preserving Our Heritage, the Industrial Base Securing Our Future*. Accession Number ADA202840, Washington, D.C., July 15, 1988.

R. B. Costello and M. Ernst, *Regaining U.S. Manufacturing Leadership*, Hudson Institute, 1992.

J. Cunningham, O. Fiume, and L. T. White, *Real Numbers, Management Accounting in a Lean Organization*, Managing Time Press, 2003.

Deloitte, *Mastering Complexity in Global Manufacturing: Powering Profits and Growth Through Value Chain Synchronization*, New York and London, Deloitte Research, 2003.

W. E. Deming, *Out of the Crisis*, MIT, Center for Advanced Engineering Study, hardcover edition, 1982.

Department of Defense, *DoD Total Quality Management Master Plan*, Washington, D.C., Accession Number ADA355612, Aug. 1988.

Department of Defense, *Instruction number 5000.02, Operation of the Defense Acquisition System*, December 8, 2008

DoDAF (Architecture Framework), V2.0, May 28, 2009 (http://cio-nii.defense.gov/sites/dodaf20/DoDAF2–0_web.pdf, last accessed April 12, 2011).

N. Egbert, P. McCoy, D. Schwerin, et al., *Design for Process Excellence—Ensuring Timely and Cost-Effective Solutions*. 26th International Congress of the Aeronautical Sciences, Anchorage, AK, 2008.

M. L. Emiliani, *Improving Business School Courses by Applying Lean Principles and Practices*. Quality Assurance in Education, 2004, 12(4), pp. 175–187.

G. Fanmuy, verbal presentation, *Lean Systems Engineering Working Group Meeting*. INCOSE 2010 International Symposium, July 10–15, 2010, Chicago.

W. Ferster, *U.S. will Try to Destroy Crippled Satellite. Space News*, February 14 2008, (http://www.space.com/4977-destroy-crippled-satellite.html, last accessed April 12, 2011).

GAO, *Defense Acquisitions: Assessments of Selected Weapon Programs*, GAO-07-4065SP, Washington, D.C., March, 2007.

GAO, *Best Practices: Increased Focus on Requirements and Oversight Needed to Improve DOD's Acquisition Environment and Weapon System Quality*, GAO-08–294, Washington, D.C., Feb. 2008a.

GAO, *Space Acquisitions: Major Space Programs Still at Risk for Cost and Schedule Increases*, GAO-08-552T, Washington, D.C., Mar. 2008b.

J. H. Gittell, *The Southwest Airlines Way*, McGraw Hill, New York, NY, 2003.

E. M. Goldratt, *Critical Chain*, North River Press, Great Barrington, MA, 1997.

M. Graban, *Lean Hospitals*, Productivity Press, London, 2008.

A. C. Haggerty, *The F/A-18E/F Super Hornet as a Case Study in "Value Based" Systems Engineering*, INCOSE 2004, Toulouse, France, June 20–24, 2004.

A. C. Haggerty and E. M. Murman, *Evidence of Lean Engineering in Aircraft Programs*. 25th International Congress of the Aeronautical Sciences, Hamburg, Germany, Sept, 2006.

M. Harry, R. Schroeder, Six Sigma: The Breakthrough Management Strategy Revolutionizing The World's Top Corporations. Currency Doubleday, 2000.

HBR, Ed., *Manufacturing Excellence at Toyota*, Harvard Business School Series, paperback, 2008.

E. Hermann, *Shinichi Suzuki: The Man and His Philosophy*, Summi-Birchard, 1981.

C. Hernandez, *Challenges and Benefits to the Implementation of IPTs on Large Military Procurements*, SM Thesis, MIT Sloan School, Cambridge, MA, June 1995.

D. Hitchens, *Systems Engineering: A 21st Century Systems Methodology*, Wiley, London, 2007.

J. Horejsi, Col., Chief Engineer Ret., USAF Space and Missile Command, Los Angeles, private communication, 2009.

E. Honour, *Systems Engineering Return on Investment*, Paper 11.4.2, INCOSE 2010 International Symposium, Chicago, July 10–15, 2010.

House of Representatives, Ninety-Ninth Congress, Investigation of The Challenger Accident, Report Of The Committee on Science And Technology, 64–4200, OCT. 29, 1986.

INCOSE Systems Engineering Handbook, v.3.1, January 2007.

INCOSE Systems Engineering Handbook, v.3.2, January 2010.

INCOSE website: http://www.incose.org/about/index.aspx., 2010.

INCOSE LSE WG, Lean Systems Engineering Working Group website, 2008, http://www.incose.org/practice/techactivities/wg/leansewg.

Integrated Defense Acquisition, Technology, and Logistics Life Cycle Management Framework, DoD, May 12, 2003.

Jacobson, C. T., TRW 1901–2001, TRW Inc., Cleveland, Ohio, 2001.

N. R. Joglekar and D. E. Whitney, *Where does time go? Design automation usage patterns during complex electro-mechanical product development*, LAI Product Development, Winter 2000 Workshop, Jan. 26–28, 2000, Fulsom, CA.

C. M. Jones, *Leading Rockwell Collins Lean Transformation*, LAI Keynote Talk, April 9, 2006, http://mitworld.mit.edu/video/378/.

S. B. Johnson, *The Secret of Apollo, Systems Management in American and European Space Programs*, John Hopkins, New Series in NASA History, 2002.

A. K. Kamrani, S. M. Salhieh, *Product Design for Modularity*, Kluwer Academic Publishers, Norwell, MA, 2002.

M. N. Kennedy, *Product Development for the Lean Enterprise*, The Oklea Press, Richmond, VA, 2003.

R. Kerber and V. Vitto, Co-Chairs of Defense Science Board, in *MEMORANDUM for Chairman, Defense Science Board, OSD*, March 2009.

LAI (Lean Advancement Initiative) website, MIT, http://lean.mit.edu/index.php?option=com_content&task=view&id=395&Itemid=336.

LAI Lean Academy® Course, Massachusetts Institute of Technology, http://ocw.mit.edu/ OcwWeb/Aeronautics-and-Astronautics/16-660January-IAP-2008/CourseHome/index. htm, 2008.

Lean Panel: *Lean-Who Can Afford to Ignore It?* D. Secor moderating, N. Malateaux, B. W. Oppenheim, H. G. Sillitto, A. Zonnenshain, INCOSE 2010 International Symposium, Chicago, July 10–15, 2010.

LEI ed., *Reflections On Lean*, Lean Enterprise Institute, Boston, 2007.

LEM, Lean Enterprise Model, Massachusetts Institute of Technology, LAI, 1996, http://lean.mit.edu/index.php?option=com_content&task=view&id=349&Itemid=303.

D. Lempia, *Using Lean Principles and MBE in Design and Development of Avionics Equipment at Rockwell Collins*. 26th International Congress of the Aeronautical Sciences, Anchorage, AK, Sept 16, 2008.

LESAT, Lean Enterprise Self Assessment Tool, Version 1.0, Lean Advancement Initiative, August 2001.

R. Leopold, *The Iridium Story: An Engineer's Eclectic Journey*, Minta Martin Lecture, MIT Department of Aeronautics and Astronautics, Apr. 23, 2004.

J. K. Liker, *The Toyota Way, 14 Management Principles*, McGraw Hill, New York, 2004.

J. K. Liker and T. N. Ogden, *Toyota Under Fire, How Toyota Faced the Challenges of the Recall and the Recession to Come Out Stronger*, McGraw Hill, New York, 2011.

J. K. Liker, D. K. Sobek II, A. C. Ward, and J. J. Cristiano, *Involving Suppliers in Product Development in the US and Japan: Evidence for Set-Based Concurrent Engineering*. IEEE Transactions in Engineering Management, Vol. 43, No. 2, May, 1996.

H. L. McManus, *Product Development Value Stream Mapping Manual*, LAI Release Beta, Massachusetts Institute of Technology, LAI, April 2004.

McManus, H, Haggerty, A., Murman, E., *Lean Engineering: A Framework for Doing the Right Job Right*. The Aeronautical Journal, Vol. 111, No. 1116, February 2007, pp. 105–114.

A. Mauritz, *EADS Innovation Works*, Academic Forum Panel on Current and Future Trends, Session IV, INCOSE 2010 International Symposium, Chicago, July 10–15, 2010.

R. L. Millard, *Value Stream Analysis and Mapping for Product Development*, Master's Thesis in Aeronautics and Astronautics, Massachusetts Institute of Technology, June 2001.

M. J. Morgan, and J. K. Liker, *Toyota Product Development System*. Productivity Press, New York, 2006.

E. M. Murman, *Lean Systems Engineering I, II, Lecture notes*, Massachusetts Institute of Technology, Course 16.885J, Fall 2003.

E. M. Murman, *Lean Aerospace Engineering*, Littlewood Lecture AIAA-2008-4, Jan. 2008.

E. M. Murman, T. Allen, K. Bozdogan, J. Cutcher-Gershenfeld, H. McManus, D. Nightingale, E. Rebentisch, T. Shields, F. Stahl, M. Walton, J. Warmkessel, S. Weiss, and S. Widnall, Lean Enterprise Value: Insights from MIT's Lean Aerospace Initiative, Palgrave, Hampshire, 2002.

NASA, Columbia Accident Investigation, Final Report, CAIB PA 40-03, August 26, 2003.

NASA Pilot Benchmarking Initiative: Exploring Design Excellence Leading to Improved Safety and Reliability, Oct. 2007.

NASA, *NASA's Toyota Study Released by Dept. of Transportation*, NASA News, February 8, 2011. (http://www.nasa.gov/topics/nasalife/features/nesc-toyota-study.html, last viewed on April 14, 2011).

NBC, *If Japan can why can't we?* TV program, June 24, 1980.

M. Nizza, *A Successful Mission*, New York Times, Feb. 25, 2008.

J. Oehman, E. Rebenstish, *Risk Management in Lean PD*, LAI Paper Series "Lean Product Development for Practitioners", March 2010.

J. Oehman, E. Rebenstish, *Waste in Lean Product Development*, LAI Paper Series "Lean Product Development for Practitioners", July 2010.

T. Ohno, *The Toyota Production System: Beyond Large Scales Production*, Productivity Press, Portland, Oregon, 1988.

B. W. Oppenheim, Lean Product Development Flow, *Journal of Systems Engineering*, Vol. 7, No. 4, 2004.

B. W. Oppenheim, *Lean as a Way of Thinking*, cover-story interview with, Quality Management Journal *(in Polish)*, Nr. 3, Vol. 5, 2006.

R. O'Rourke, *Coast Guard Deepwater Acquisition Programs: Background, Oversight Issues, and Options for Congress*, Congressional Research Service, Order Code RL33753, Updated October 9, 2008.

S. M. Paton, *Is TQM Dead?* Quality Digest, April 1994.

E. Plotkin, *Lean Enablers for Systems Engineering*, Master's Thesis in Mechanical Engineering, Loyola Marymount University, Sep. 2008.

E. Rebentisch, *Lean Product Development*, Massachusetts Institute of Technology, LAI lecture, Oct. 5, 2005.

E. Rebentisch, H. McManus, *Tutorial on Lean PD*, Massachusetts Institute of Technology, Lean Advancement Initiative, 2007.

E. Rebentisch, D. Rhodes, E. Murman, *Lean Systems Engineering: Research Initiatives in Support of a New Paradigm*, Conference on Systems Engineering Research, University of Southern California, 122, April 2004.

D. G. Reinertsen, *The Principles of Product Development Flow*, 2nd Generation LPD, Celeritas Publ., 2009.

D. Rhodes, C. Dagli, A. Haggerty, R. Jain, E. Rebentisch, *Panel on Lean Systems Engineering*, INCOSE 2004, Toulouse, France, June 20–24, 2004.

B. R. Rich, L. Janos, *Skunk Works: A Personal Memoir of My Years at Lockheed*, Little, Brown and Company, Canada, 1994.

Rockwell Collins, *Lean Benchmarking Event*, Rockwell Collins, 2007.

T. Rockwell, *The Rickover Effect: The Inside Story of How Adm. Hyman Rickover Built the Nuclear Navy*, John Wiley & Sons, New York, 1995.

W. H. Schmidt, J. P. Finnigan, Visit Amazon's Warren H. Schmidt Pagesearch result Learn about Author Central, *Race Without a Finish Line: America's Quest for Total Quality*. Jossey Bass Business and Management Series, 1992.

D. Secor, Lean SE Working Group Meeting, INCOSE 2010 International Symposium, Chicago, July 10–15, 2010.

R. A. Slack, *Application of Lean Principles to the Military Aerospace Product Development Process*, Masters Thesis in Engineering and Management, Massachusetts Institute of Technology, Dec. 1998.

D. K., Sobek II, *Principles Shape Product Development Systems: A Toyota-Chrysler Comparison*, Ph.D. Thesis, U. Michigan, 1997a.

D. K. Sobek II, *Toyota's Product Development Process*, a case study in M. Fleischer and J. Liker Concurrent Engineering Effectiveness: Concepts and Methods, Hanson-Gardner Publishers,; pp. 461–480, 1997b.

D. K., Sobek II, J. K. Liker and A. C. Ward, *Another Look at Toyota's Integrated Product Development*, Harvard Business Review, Vol. 76, No. 4, July-August, 1998; pp. 36–49.

D. K. Sobek II, A. C. Ward, and J. K. Liker, *Toyota's Principles of Set-Based Concurrent Engineering*, Sloan Management Review, Vol. 40, No. 2, Winter, 1999; pp. 67–83.

S. Spear and H. K. Bowen, *Decoding the DNA of TPS*, Harvard Bus Rev 77(5) (1999), 99–109.

A. Stanke, *A Framework for Achieving Lifecycle Value in Product Development*, SM Thesis in Aeronautics and Astronautics, Massachusetts Institute of Technology, 2001.

Y. Sugimori, K. Kisunoki, F. Cho, S. Uchikawa, *Toyota Production System and Kanban Systems-Materialization of Just-In-Time and Respect-For-Human Systems*, International Journal of Production Research, Vol. 15, No. 6 (1977), pp. 553–64.

K. T. Ulrich and S. D. Eppinger, *Product Design and Development*, 4th ed, New York, McGraw-Hill, 2008.

A. C. Ward, *Lean Product and Process Development*, Lean Enterprise Institute, Cambridge, MA, Mar 2007.

A. C. Ward, J. K. Liker, J. J. Cristiano, and D. K. Sobek II, *The Second Toyota Paradox: How Delaying Decisions Can Make Better Cars Faster*, Sloan Management Review, Vol. 36, No. 3, Spring 1995a.

A. C. Ward, D. K. Sobek II, J. J. Cristiano, and J. K. Liker, *Toyota, Concurrent Engineering, and Set-Based Design*, a chapter of Liker, et al. (ed.), *Engineered in Japan: Japanese Technology Management Practices*, Oxford Press, New York, 1995b.

J. Warmkessel, *Lean Engineering*, Lean Aerospace Initiative, MIT, http://lean.mit.edu, 2002.

G. Warwick, *Opening doors: Car maker Honda's aircraft research and development facility gears up for the HondaJet*, Flight International, Dec. 1, 2007.

L. Webb, *Knowledge Management for Through Life Support*, PhD Thesis in progress, private communication, RMIT University, Australia, 2008.

I. Wedgewood, *Lean Six Sigma, A Practitioner's Guide*, Prentice Hall, Upper Saddle River, NJ, 2007.

W. L. Wilson, S. Malik, B. Irwin, *Enterprise Engineering*, Lockheed Martin Center for Enterprise Engineering, 2010.

I. R. Winner, P. J. Pennell, E. H. Bertrand, M. M. Slusarczuk, *The Role of Concurrent Engineering in Weapons System Acquisition*, Institute for Defense Analysis, Report-R-338, 1988.

J. P. Womack, D. T. Jones, *Lean Thinking*, Simon & Shuster, London, 1996.

J. P. Womack, D. T. Jones, D. Roos, *The Machine That Changed The World, The Story of Lean Production*, The MIT International Motor Vehicle Program, Harper-Perennial, 1990.

K. Yang and B. El-Haik, *Design for Six Sigma: A Roadmap for Product Development*, Mc-Graw Hill, New York, 2003.

J. Young, *Memo to the Secretary of Defense*, Department of Defense, January 30, 2009.

T. Young, *Report on Mars Program Failure*, U.S. Congress, Science Committee, April 12, 2000.

Appendix **1**

INCOSE Web Page with LEfSE

The public web page[1] www.incose.org/practice/techactivities/wg/leansewg/on the International Council on Systems Engineering (INCOSE) web site contains the following products related to the Lean Enablers for Systems Engineering (EfSE):

- Photo and text of the 2010 INCOSE Best Product Award, and logo of the Shingo Award.
- Power point presentation Lean Enablers for Systems Engineering, Version 1 (pdf, 0.764 MB)
- Three articles:
  - Lean Enablers for Systems Engineering, B.W. Oppenheim, E. Murman, D. Secor, full-length article, Journal of SE, Jan. 2010 (online version).
  - Lean Enablers for Systems Engineering, five-page summary article from Cross Talk (pdf, 0.798 MB).
  - Lean Enablers for Systems Engineering, short article from INCOSE INSIGHT (pdf, 0.524 MB).
- Summary text about Lean SE (see INCOSE SE Handbook v. 3.2).
- Lean SE Brochure (trifold, letter format, pdf, 0.859 MB) created in cooperation with INCOSE UK.

---

[1] An easy to remember access to the website is: www.INCOSE.org, then click on Working Groups, and then on Lean Systems Engineering.

- LEfSE Quick Reference Guide (8 pages, pdf, 0.268 MB) created in cooperation with INCOSE UK.
- Three-piece video with lecture by B. Oppenheim to Booze, Allen, Hamilton (1 hr 37 min total), with slides integrated.
- A list of workshops, tutorial and lectures delivered on Lean Enablers for SE.

The page also contains the Lean Systems Engineering Charter, leadership and contact information, recommended reading list, definitions, information about meetings and conferences, and other useful information.

Appendix 2

Mapping of LEfSE onto INCOSE SE Processes

Besides the SE Processes, at the time of this writing several other frameworks for defining program life cycle exist: Enterprise Architecting and Engineering [e.g. Wilson et al., 2010]; DoD Architecture Framework [DODAF, 2009], CMMI [2010]; Integrated Defense Acquisition, Technology, and Logistics Life Cycle Management Framework [DoD, 2003], and several others issued by the individual military branches. Most are mandatory in DoD contracts, while others are not. This is all in addition to about 60 Military Standards that have been mandated for most DoD programs. Usually, these frameworks compete with others for the top position in the program management hierarchy, decision priority and resource allocation. Most of them contain excellent ideas which are helpful in program management. However, developed in uncoordinated manner by competing organizations they struggle for the "top position" adding immeasurable amount of waste and bureaucracy. As a result, the number of requirements defined for programs grow into thousands. The programs struggling to satisfy this massive number of competing frameworks, standards, and requirements waste a significant portion of program budget and schedule, leaving fewer resources for the most important part of the program: real engineering. In the opinion of this author, this situation is unsustainable. It makes U.S. programs less and less competitive globally, vastly more expensive, longer, and producing imperfect quality. A most serious effort is recommended to integrate all frameworks into a consistent and lean body of knowledge, keeping only those elements which truly add value and eliminate waste. At that time, Lean Enablers for SE should also be mapped accordingly. Until that time, LEfSE mapping is limited to the INCOSE SE processes.

Lean for Systems Engineering with Lean Enablers for Systems Engineering,
First Edition. Bohdan W. Oppenheim.
© 2011 John Wiley & Sons, Inc. Published 2011 by John Wiley & Sons, Inc.

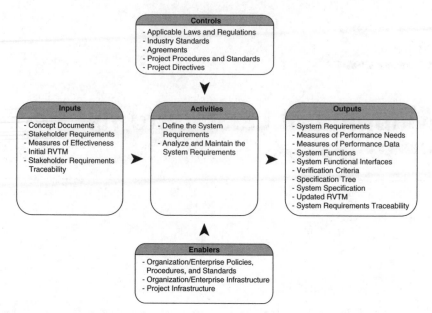

Figure App. 2.1 Context Diagram for Implementation Process. [INCOSE, 2010, Fig. 4.12]

Version 3.2 of INCOSE SE Handbook [2010] partitions Systems Engineering into 26 processes. This Appendix presents a mapping of the 147 LEfSE subenablers onto those 26 processes.[1]

The Handbook illustrates each process with a context diagram. Figure App. 2.1 illustrates a sample context diagram for the Implementation Process. All SE processes are modeled as shown in Figure App. 2.1, with five boxes titled: Inputs, Activities, Outputs, Controls, and General Enablers.

The boxes labeled General Enablers in different diagrams include various combinations of the following bullets:

- Organizational/Enterprise Policies, Procedures, and Standards
- Organizational/Enterprise Infrastructure
- Project Infrastructure
- Implementation Enabling System

These General Enablers should not be confused with LEfSE enablers and subenablers presented in this book. Obviously, INCOSE General Enablers are not focused

---

[1]The mapping was initiated by INCOSE leadership. The version listed in this Appendix has already been reviewed, approved with minor corrections by the INCOSE Lean SE Working Group during the January 29, 2011 International Workshop, and proposed for insertion into the next edition of the INCOSE SE Handbook. However, the Handbook Committee of INCOSE has not yet approved this text. Until it does, the text should be regarded only as a draft. The text has been developed from the M.Sc. Capstone Project (of the same title as the present Appendix) performed by Dana Makiewicz, graduate student of Systems Engineering at Loyola Marymount University in Los Angeles in the Fall of 2010.

on Lean, and are defined at much higher level than the detailed and actionable LEfSE subenablers.

Because of large numbers of subenablers involved, they are listed in text (below) rather than in boxes in context diagrams. The Lean SE Working group of INCOSE recommended to the Handbook Committee that an Appendix titled "Lean Enablers for SE Processes" with the subenabler text be added to a future version of the INCOSE SE Handbook. It was also recommended to add a small symbolic sixth box titled "Lean Enablers and Subenablers in Appendix x" to each context diagram in the Handbook, to remind readers that Lean subenablers should be applied to each process, and they can be found in Appendix x.

The mapping of 147 LEfSE subenablers onto the 26 INCOSE processes was done iteratively, by trial and error. Often, this was difficult because a large fraction of Lean subenablers apply to more than one process (confirming what we discovered early in the original development of LEfSE). Nevertheless, we placed each Lean subenabler in only one process which was judged the most appropriate for it from the point of view of implementation. The largest group of 47 subenablers was judged to apply to all INCOSE processes, and those are listed below under a special heading "*All Processes*". They represent powerful Lean ideas on how to improve many aspects of work in all processes.

We decided to define a new process, termed *Enterprise Preparation Process (EPP)*. It lists those 10 Lean subenablers which benefit all present and future programs in the Enterprise, and therefore should be implemented at the Enterprise (corporate) rather than program level, if possible.

Twelve processes ended up with no Lean subenabler. It is explained below that these processes benefit indirectly from other subenablers. Box App. 2.1 summarizes the mapping results.

The data in Box App. 2.1 inspires several interesting observations, as follows:

- The largest category is All Processes with 47 subenablers. This is not surprising. These subenablers address the critical aspects of SE which are often ignored in traditional programs and in SE handbooks, and which flow naturally from Lean Thinking: excellent coordination and communication, alignment for customer value, teamwork, better interactions between stakeholders, emphasis on performing work *right the first time*, excellent interpersonal relations and human habits, etc.
- The next in size is the Project Planning Process with 32 subenablers. This is consistent with a strong focus of LEfSE on improving front-end activities of programs: better preparations, better planning for value capture, better planning of program, planning for best communication and coordination means, better frontloading, stronger integration of SE and PD, and better human relations among stakeholders. Again, these practices are often performed in traditional programs not as well as they could be.
- Fourteen SE processes have between 1 and 10 Lean subenablers. These are manageable numbers for implementations.

Box App. 2.1 Summary of LEfSE Subenablers Mapping onto INCOSE SE Processes

INCOSE SE Process	Number of LEfSE Subenablers Mapped
All Processes (in addition to those listed below)	47
Enterprise Preparation Process	10
4.1 Stakeholder Requirements Definition Process	2
4.2 Requirements Analysis Process	0
4.3 Architectural Design Process	7
4.4 Implementation Process	4
4.5 Integration Process	0
4.6 Verification Process	0
4.7 Transition Process	0
4.8 Validation Process	0
4.9 Operation Process	0
4.10 Maintenance Process	0
4.11 Disposal Process	0
5.1 Project Planning Process	32
5.2 Project Assessment and Control Process	0
5.3 Decision Management Process	2
5.4 Risk Management Process	2
5.5 Configuration Management Process	1
5.6 Information Management Process	6
5.7 Measurement Process	5
6.1 Acquisition Process	0
6.2 Supply Process	7
7.1 Life Cycle Model Management Process	0
7.2 Infrastructure Management Process	0
7.3 Project Portfolio Management Process	3
7.4 Human Resource Management Process	10
7.5 Quality Management Process	9
8.1 Tailoring Process	1

- Twelve SE Processes indicate zero dedicated Lean subenablers: Requirements Analysis, Integration, Verification, Transition, Validation, Operations, Maintenance, Disposal, Project Assessment and Control, Acquisition, Life Cycle Model Management, and Infrastructure Management. This is not an indication that these twelve processes need no Lean wisdom. Instead, the processes are being improved indirectly, by applying Lean wisdom to the front-end processes when most of the critical decisions are made (enterprise and program preparations, program planning, value capture, design frontloading, best engineering practices, implementation, quality, and management). In addition, the 47 Lean subenablers listed under ALL Processes will improve the twelve processes too.
- Note: the sum of subenablers in Box App. 2.1 is 148 rather than 147 used elsewhere in this book. This is because the important enabler 6.5 which comprises no subenablers, has been included.

MAPPING OF LEfSE SUBENABLERS ONTO 26 INCOSE SE PROCESSES

The following is a list of the LEfSE subenablers[2] mapped into individual INCOSE process, starting with the added category called "All Processes", and the added Enterprise Preparation Process, and then following the order of processes in the INCOSE SE Handbook. Note: The bold and italic markings used elsewhere in this book have been suppressed from the following list, to conform to the INCOSE SE Handbook style.

All Processes (These subenablers apply to all SE processes)

1.2.1. Define value as the outcome of an activity that satisfies at least three conditions:

   a. The external customer is willing to pay for value.

   b. Transforms information or material or reduces uncertainty.

   c. Provides specified performance right the first time.

1.2.2. Define value-added in terms of value to the customer and his needs.

1.2.5. Do not ignore potential conflicts with other stakeholder values, and seek consensus.

1.3.1. Everyone involved in the program must have a customer-first spirit.

1.3.2. Establish frequent and effective interaction with internal and external customers.

2.2.6. Scrutinize every step to ensure it adds value, and plan nothing because "it has always been done."

---

[2]The version shown here and proposed to INCOSE is slightly different from the text in this book: abbreviations are spelled out and bold and italic fonts eliminated.

3.2.1. Since formal written requirements are rarely enough, allow for follow up verbal clarification of context and need, without allowing requirements creep.

3.2.6. To align stakeholders, identify a small number of goals and objectives that articulate what the program is set up to do, how it will do it, and what the success criteria will be, and repeat this process consistently and often.

3.4.2. SE to regard all other engineers as their partners and internal customers, and vice-versa.

3.5.1. Capture and absorb lessons learned from almost all programs: "never enough coordination and communication."

3.5.4. Use frequent, timely, open and honest communication.

3.5.5. Promote direct informal communications immediately as needed.

3.5.6. Use concise one-page electronic forms (e.g., Toyota's A3 form) rather than verbose unstructured memos to communicate, and keep detailed working data as backup.

3.6.1. Use formal frequent comprehensive integrative events in addition to programmatic reviews of the program.

    a. Question everything with multiple "whys."

    b. Align process flow to decision flow.

    c. Resolve all issues as they occur in frequent integrative events.

    d. Discuss tradeoffs and options.

3.6.2. Be willing to challenge the customer's assumptions on technical and meritocratic grounds, and to maximize program stability, relying on technical expertise.

3.6.3. Minimize handoffs to avoid rework.

3.6.4. Optimize human resources when allocating value-added (VA) and required-non-value-added (RNVA) tasks:

    a. Use engineers to do VA engineering.

    b. When engineers are not absolutely required, use non-engineers to do RNVA (administration, project management, coasting, metrics, program, etc.)

3.6.5. Ensure the use of the same measurement standards and database commonality.

3.6.6. Ensure that both data deliverers and receivers understand the mutual needs and expectations.

3.7.1. Make work progress visible and easy to understand to all, including external customer.

3.7.2. Utilize Visual Controls in public spaces for best visibility (avoid computer screens).

3.7.3. Develop a system making imperfections and delays visible to all.

3.7.4. Use traffic light system (green, yellow, red) to report task status visually (good, warning, critical) and make certain problems are not concealed.

4.2.2. Promote the culture in which engineers pull knowledge as they need it and limit the supply of information to only genuine users.

4.2.4. Train the team to recognize who the internal customer (Receiver) is for every task as well as the supplier (Giver) to each task. Use a SIPOC (supplier, inputs, process, outputs, customer) model to better understand the value stream.

4.2.5. Stay connected to the internal customer during the task execution.

4.2.7. Promote effective real time direct communication between each Giver and Receiver in the value flow.

4.2.8. Develop Giver-Receiver relationships based on mutual trust and respect.

4.2.9. When pulling work, use customer value to delineate value adding work from waste.

5.2.2. Promote excellence under "normal" circumstances instead of hero-behavior in "crisis" situations.

5.2.3. Use and communicate failures as opportunities for learning emphasizing process and not people problems.

5.2.4. Treat any imperfection as an opportunity for immediate improvement and lesson to be learned, and practice frequent reviews of lessons learned.

5.2.5. Maintain a consistent disciplined approach to engineering.

5.3.2. Create mechanisms to capture, communicate, and apply experience-generated learning and checklists.

5.3.4. Identify best practices through benchmarking and professional literature.

6.2.3. Promote excellent human relations: trust, respect, honesty, empowerment, teamwork, stability, motivation, and drive for excellence.

6.2.5. Promote direct human communication.

6.2.9. Eliminate fear and promote conflict resolution at the lowest level.

6.2.10. Keep management decisions crystal clear but also promote and reward the bottom-up culture of continuous improvement, human creativity, and entrepreneurship.

6.2.11. Do not manage from cubicle; go to the spot and see for yourself.

6.2.12. Within program policy and within their area of work, empower people to accept responsibility by promoting the motto "ask for forgiveness rather than ask for permission."

6.2.13. Build a culture of mutual support. There is no shame asking for help.

6.4.3. Provide knowledge experts as resources and for mentoring.

6.4.4. Pursue the most powerful competitive weapon: the ability to learn rapidly and continuously improve.

6.4.5. Value people, through mutual respect and appreciation, for the skills they contribute to the program.

6.4.8. Immediately organize a quick training in any new standard.

  6.5. Treat People as Most Valued Assets, not as Commodities.

Enterprise Preparation Process

5.4.3. Promote good coordination and communications skills with training and mentoring.

5.5.1. The Chief Engineer role to be responsible, with authority and account-ability for the program technical success.

5.5.2. Have the Chief Engineer role lead both the product and people integration.

5.5.3. Have the Chief Engineer role lead through personal influence, technical know how, and authority over product development decisions.

5.5.4. Groom an exceptional Chief Engineer role with the skills to lead the development, the people, and assure program success.

5.5.5. If Program Manager and Chief Engineer role are two separate individuals (required by contract or organizational practice), co-locate both to enable constant close coordination.

5.6.1. Promote design standardization with engineering checklists, standard architecture, modularization, busses, and platforms.

5.6.2. Promote process standardization in development, management, and man-ufacturing.

6.3.1. Establish and support Communities of Practice.

6.4.7. Develop Standards paying attention to human factors, including reading and perception abilities.

4.1 Stakeholder Requirements Definition Process

3.2.2. Create effective channels for clarification of requirements (possibly involve customer participation in development Integrated Product Teams (IPTs)).

3.2.3. Listen for and capture unspoken customer requirements.

4.2 Requirements Analysis Process

No subenablers mapped onto this process.

4.3 Architectural Design Process

1.3.3. Pursue an architecture that captures customer requirements clearly and can be adaptive to changes.

2.3.2. Explore trade space and margins fully before focusing on a point design and too small margins.

2.4.2. Insist that a module proposed for use is robust before using it.

3.2.4. Use architectural methods and modeling for system representations (3D integrated Computer-Aided Engineering toolset, mockups, prototypes, models, simulations, and software design tools) that allow interactions with customers as the best means of drawing out customer requirements.

3.2.5. Fail early-fail often through rapid learning techniques (prototyping, tests, digital preassembly, spiral development, models, and simulation).

3.3.1. Explore multiple concepts, architectures and designs early.

3.3.2. Explore constraints and perform real trades before converging on a point design.

3.3.3. Use a clear architectural description of the agreed solution to plan a coherent program, engineering and commercial structures.

4.4 Implementation Process

1.2.6 Explain customer culture to Program employees, i.e. the value system, approach, attitude, expectations, and issues.

3.3.4 All other things being equal, select the simplest solution. ("Any fool can make anything complex but it takes a genius and courage to create a simple solution" —Albert Einstein.)

4.2.1 Let information needs pull the necessary work activities.

4.5 Integration Process

No subenablers mapped onto this process.

4.6 Verification Process

No subenablers mapped onto this process.

4.7 Transition Process

No subenablers mapped onto this process.

4.8 Validation Process

No subenablers mapped onto this process.

4.9 Operation Process

No subenablers mapped onto this process.

4.10 Maintenance Process

No subenablers mapped onto this process.

4.11 Disposal Process

No subenablers mapped onto this process.

5.1 Project Planning Process

1.2.3. Develop a robust process to capture, develop, and disseminate customer value with extreme clarity.

1.2.4. Develop an agile process to anticipate, accommodate and communicate changing customer requirements.

1.3.4. Establish a plan that delineates the artifacts and interactions that provide the best means for drawing out customer requirements.

2.2.1. Develop and execute clear communication plan that covers entire value stream and stakeholders.

2.2.2. Have cross functional stakeholders work together to build the agreed value stream.

2.2.3. Create a plan where both Systems Engineering (SE) and other Product Development (PD) activities are appropriately integrated.

2.2.4. Maximize co-location opportunities for SE and [other] PD planning.

2.2.7. Carefully plan for precedence of both SE and PD tasks (which task to feed what other tasks with what data and when), understanding task dependencies and parent-child relationships.

2.2.8. Maximize concurrency of SE and other PD Tasks.

2.2.9. Synchronize work flow activities using scheduling across functions, and even more detailed scheduling within functions.

2.2.10. For every action, define who is responsible, approving, supporting, and informing (RASI), using a standard and effective tool, paying attention to precedence of tasks.

2.2.11. Plan for level workflow and with precision to enable schedule adherence and drive out arrival time variation.

2.2.12. Plan below full capacity to enable flow of work without accumulation of variability, and permit scheduling flexibility in work loading, i.e., have appropriate contingencies and schedule buffers.

2.2.13. Plan to use visual methods wherever possible to communicate schedules, workloads, changes to customer requirements, and other information.

2.3.1. Plan to utilize cross-functional teams made up of the most experienced and compatible people at the start of the project to look at a broad range of solution sets.

2.3.4. Plan early for consistent robustness and "right the first time" under "normal" circumstances instead of hero behavior in later "crisis" situations.

2.5.3. Plan to include and manage the major suppliers as a part of your team.

3.4.1. Promote maximum seamless teaming of Systems Engineers and other Product Development engineers.

3.4.3. Maintain team continuity between phases to maximize experiential learning.

3.4.4. Plan for maximum continuity of Systems Engineering staff during the program.

3.5.2. Maximize coordination of effort and flow (one of the main responsibilities of Lean SE).

3.8.1. Use Lean tools to promote the flow of information and minimize handoffs: small batch size of information, small takt times, wide communication bandwidth, standardization, work cells, training.

4.2.3. Understand the Value Stream Flow.

4.2.6. Avoid rework by coordinating task requirements with internal customer for every non-routine task.

5.2.8. Use a balanced matrix/project organizational approach avoiding extremes: territorial functional organizations with isolated technical specialists, and all-powerful Integrated Product Teams (IPTs) separated from functional expertise and standardization.

5.4.1. Develop a plan and train the entire program team in communications and coordination methods at the program beginning.

5.4.4. Publish instructions for e-mail distributions and electronic communications.

5.4.6. Publish a directory of the entire program team and provide training to new hires on how to locate the needed nodes of knowledge.

6.2.1. Create a vision which draws and inspires the best people.

6.2.2. Invest in people selection and development to promote enterprise and program excellence.

6.2.14. Prefer physical team co-location to the virtual co-location.

6.4.6. Capture learning to stabilize the program when people transfer elsewhere or leave.

5.2 Project Assessment and Control Process

No subenablers mapped onto this process.

5.3 Decision Management Process

3.5.7. Report cross-functional issues to be resolved on concise standard one-page forms to Chief's office in real time for his/her prompt resolution.

6.2.8. Flow down responsibility, authority, and accountability (RAA) to allow decision making at lowest appropriate level.

5.4 Risk Management Process

2.3.3. Anticipate and plan to resolve as many downstream issues and risks as early as possible to prevent downstream problems.

2.4.3. Remove show-stopping research/unproven technology from critical path, staff with experts, and include in the Risk Mitigation Plan.

5.5 Configuration Management Process

5.4.5. Publish instructions for artifact content and data storage—central capture versus local storage, and for paper versus electronic—balancing between excessive bureaucracy and the need for traceability.

5.6 Information Management Process

3.8.2. Use minimum number of tools and make common wherever possible.

3.8.3. Minimize the number of the software revision updates and centrally control the update releases to prevent information churning.

3.8.4. Adapt the technology to fit the people and process.

3.8.5. Avoid excessively complex "monument" tools.

5.4.7. Ensure timely and efficient access to centralized data.

5.4.8. Develop an effective body of knowledge that is historical, searchable, shared by team, and knowledge management strategy to enable the sharing of data and information within the enterprise.

5.7 Measurement Process

2.6.1. Use leading indicators to enable action before waste occurs.

2.6.2. Focus metrics around customer value, not profits.

2.6.3. Use only few simple and easy to understand metrics and share them frequently throughout the enterprise.

2.6.4. Use metrics structured to motivate the right behavior.

2.6.5. Use only those metrics that meet a stated need or objective.

6.1 Acquisition Process

No subenablers mapped onto this process.

6.2 Supply Process

2.5.1. Select suppliers who are technically and culturally compatible.

2.5.2. Strive to develop seamless partnership between suppliers and the product development team.

2.5.4. Have the suppliers brief the design team on current and future capabilities during conceptual formation of the project.

3.3.5. Invite suppliers to make a serious contribution to SE, design and development as program trusted partner.

3.5.3. Maintain counterparts with active working relationships throughout the enterprise to facilitate efficient communication and coordination among different parts of the enterprise, and with suppliers.

3.5.8. Communicate to suppliers with crystal clarity all expectations, including the context and need, and all procedures and expectations for acceptance tests, and ensure the requirements are stable.

3.5.9. Trust engineers to communicate with suppliers' engineers directly for efficient clarification, within a framework of rules, (but watch for high risk items which must be handled at the top level).

7.1 Life Cycle Model Management Process

No subenablers mapped onto this process.

7.2 Infrastructure Management Process

No subenablers mapped onto this process.

7.3 Project Portfolio Management Process

2.4.1. Promote reuse and sharing of program assets: Utilize platforms, standards, busses, and modules of knowledge, hardware and software.

2.4.4. Defer unproven technology to future technology development efforts, or future systems.

2.4.5. Maximize opportunities for future upgrades, (e.g., reserve some volume, mass, electric power, computer power, and connector pins), even if the contract calls for only one item.

7.4 Human Resource Management Process

5.4.2. Include communication competence among the desired skills during hiring.

5.6.3. Promote standardized skill sets with careful training and mentoring, rotations, strategic assignments, and assessments of competencies.

6.2.4. Read applicant's resume carefully for both technical and nontechnical skills, and do not allow mindless computer scanning for keywords.

6.2.6. Promote and honor technical meritocracy.

6.2.7. Reward based upon team performance, and include teaming ability among the criteria for hiring and promotion.

6.3.2. Invest in Workforce Development.

6.3.3. Assure tailored lean training for all employees.

6.3.4. Give leaders at all levels in-depth lean training.

6.4.1. Perpetuate technical excellence through mentoring, training, continuing education, and other means.

6.4.2. Promote and reward continuous learning through education and experiential learning.

7.5 Quality Management Process

5.2.1. Do not ignore the basics of Quality:

   a. Build in robust quality at each step of the process, and resolve and do not pass along problems.

    b. Strive for perfection in each process step without introducing waste.

    c. Do not rely on final inspection; error proof wherever possible.

    d. If final inspection is required by contract, perfect upstream processes pursuing 100% inspection pass rate.

    e. Move final inspectors upstream to take the role of quality mentors.

    f. Apply basic PDCA method (plan, do, check, act) to problem solving.

    g. Adopt and promote a culture of stopping and permanently fixing a problem as soon as it becomes apparent.

5.2.6. Promote the idea that the system should incorporate continuous improvement in the organizational culture, but also... (continued in 5.2.7)

5.2.7. ...balance the need for excellence with avoidance of overproduction waste (pursue refinement to the point of assuring value and "right the first time", and prevent over-processing waste).

5.3.1. Maximize opportunities to make each next program better then the last.

5.3.3. Insist on workforce training of root cause and appropriate corrective action.

5.3.5. Share metrics of supplier performance back to them so they can improve.

5.7.1. Utilize and reward bottom-up suggestions for solving employee-level problems.

5.7.2. Use quick response small Kaizen teams comprised of problem stakeholders for local problems and development of standards.

5.7.3. Use the formal large Six Sigma teams for the problems which cannot be addressed by the bottom-up and Kaizen improvement systems, and do not let the Six Sigma program destroy those systems.

8.1 Tailoring Process

2.2.5. Use formal value stream mapping methods to identify and eliminate SE and other PD waste, and to tailor and scale tasks.

Author's Biography

Bohdan W. Oppenheim was born in 1948 in Warsaw, Poland. He has been living in the United States since 1971. His degrees include PhD (1980) from University of Southampton, U.K. in Ship Dynamics; Naval Architect's postgraduate degree (1974) from MIT; M.S. from Stevens Institute of Technology in New Jersey in Ocean Systems (1972); and B.S. (equivalent, 1970), from Warsaw Technical University in Mechanical Engineering and Aeronautics ("MEL").

He began working at Loyola Marymount University, Los Angeles in 1982 as Professor of Mechanical Engineering. He became a Professor of Systems Engineering in 2004. He also held the position of Graduate Director of Mechanical Engineering during 1995–2009.

He is the Founder and Co-Chair of Lean Systems Engineering Working Group of INCOSE and leader of the Prototype team developing Lean Enablers for Systems Engineering; and the recipient of INCOSE Product of the Year Award; Shingo Award in 2010; Fulbright Award, 2011. He was also awarded *Best Engineering Teacher* by the Los Angeles Council of Engineers and Scientists, 2008 and $1,922,000 in externally funded grants.

His professional experience includes serving as the Director of the U.S. Department of Energy Industrial Assessment Center (assessed 125 industrial plants for lean productivity, 2000–2007); Coordinator of Lean Advancement Initiative Educational Network and serving on the Steering Committee of the Lean Education Academic Network. Areas of specialization are lean, productivity, quality, resiliency, systems

Lean for Systems Engineering with Lean Enablers for Systems Engineering,
First Edition. Bohdan W. Oppenheim.
© 2011 John Wiley & Sons, Inc. Published 2011 by John Wiley & Sons, Inc.

engineering, and formerly included dynamics, signal processing, vessel mooring simulators, and naval architecture.

His industrial experience includes serving as Research Engineer at Aerospace Corporation (1990–1994); Northrop (1985–1990); Global Marine Development (1974–78); consultancy to: Northrop-Grumman (2007–2008), Boeing (2001–2004), Airbus (2005), Telekomunikacja Polska (2006–2008), Mars (2007–2008), and 50 other firms and governmental institutions in the U.S. and Europe.

He is a member of INCOSE, LAI EdNet, PIASA, and formerly ASEE, ASME, ISOPE, NPAJAC and SNAME and a Fellow at IAE. He has authored *POGO Oscillation Simulator for Liquid Rockets*, used by rocket industry and NASA, developed at The Aerospace Corporation. And is also an author of 21 technical journal articles, books and book chapters, and 12 non-technical articles, books and TV programs. He has presented 35 workshops, tutorials and webinars on Lean Enablers for Systems Engineering, and 13 workshops, tutorials and webinars on Lean Product Development Flow and been Guest Lecturer in Canada, France, Germany, Italy, Israel, Netherlands, Norway, Poland, Sweden, and United Kingdom. He lives in Santa Monica, California.

He has two sons, enjoys ocean sailing and is a collector of modern Polish art.

Index

Lean for Systems Engineering with Lean Enablers for Systems Engineering,
First Edition. Bohdan W. Oppenheim.
© 2011 John Wiley & Sons, Inc. Published 2011 by John Wiley & Sons, Inc.

WILEY SERIES IN SYSTEMS ENGINEERING AND MANAGEMENT

Andrew P. Sage, Editor

YACOV Y. HAIMES
Risk Modeling, Assessment, and Management, Third Edition

DENNIS M. BUEDE
The Engineering Design of Systems: Models and Methods, Second Edition

ANDREW P. SAGE and JAMES E. ARMSTRONG, Jr.
Introduction to Systems Engineering

WILLIAM B. ROUSE
Essential Challenges of Strategic Management

YEFIM FASSER and DONALD BRETTNER
Management for Quality in High-Technology Enterprises

THOMAS B. SHERIDAN
Humans and Automation: System Design and Research Issues

ALEXANDER KOSSIAKOFF and WILLIAM N. SWEET
Systems Engineering Principles and Practice

HAROLD R. BOOHER
Handbook of Human Systems Integration

JEFFREY T. POLLOCK and RALPH HODGSON
Adaptive Information: Improving Business Through Semantic Interoperability, Grid Computing, and Enterprise Integration

ALAN L. PORTER and SCOTT W. CUNNINGHAM
Tech Mining: Exploiting New Technologies for Competitive Advantage

REX BROWN
Rational Choice and Judgment: Decision Analysis for the Decider

WILLIAM B. ROUSE and KENNETH R. BOFF (editors)
Organizational Simulation

HOWARD EISNER
Managing Complex Systems: Thinking Outside the Box

STEVE BELL
Lean Enterprise Systems: Using IT for Continuous Improvement

J. JERRY KAUFMAN and ROY WOODHEAD
Stimulating Innovation in Products and Services: With Function Analysis and Mapping

WILLIAM B. ROUSE
Enterprise Tranformation: Understanding and Enabling Fundamental Change

JOHN E. GIBSON, WILLIAM T. SCHERER, and WILLAM F. GIBSON
How to Do Systems Analysis

WILLIAM F. CHRISTOPHER
Holistic Management: Managing What Matters for Company Success

WILLIAM B. ROUSE
People and Organizations: Explorations of Human-Centered Design

MO JAMSHIDI
System of Systems Engineering: Innovations for the Twenty-First Century

ANDREW P. SAGE and WILLIAM B. ROUSE
Handbook of Systems Engineering and Management, Second Edition

JOHN R. CLYMER
Simulation-Based Engineering of Complex Systems, Second Edition

KRAG BROTBY
Information Security Governance: A Practical Development and Implementation Approach

JULIAN TALBOT and MILES JAKEMAN
Security Risk Management Body of Knowledge

SCOTT JACKSON
Architecting Resilient Systems: Accident Avoidance and Survival and Recovery from Disruptions

JAMES A. GEORGE and JAMES A. RODGER
Smart Data: Enterprise Performance Optimization Strategy

YORAM KOREN
The Global Manufacturing Revolution: Product-Process-Business Integration and Reconfigurable Systems

AVNER ENGEL
Verification, Validation, and Testing of Engineered Systems

WILLIAM B. ROUSE (editor)
The Economics of Human Systems Integration: Valuation of Investments in People's Training and Education, Safety and Health, and Work Productivity

ALEXANDER KOSSIAKOFF, WILLIAM N. SWEET, SAM SEYMOUR, and STEVEN M. BIEMER
Systems Engineering Principles and Practice, Second Edition

GREGORY S. PARNELL, PATRICK J. DRISCOLL, and DALE L. HENDERSON (editors)
Decision Making in Systems Engineering and Management, Second Edition